GLOBAL CLASSIC LANDSCAPE DESIGN EXPLORATION HIGHLIGHTS

Global Classic Landscape Design Exploration Highlights

全球经典景观设计探索集锦 I

《景观设计》杂志社 编

大连理工大学出版社

图书在版编目（CIP）数据

全球经典景观设计探索集锦：全4册 /《景观设计》
杂志社编. -- 大连：大连理工大学出版社，2011.9
 ISBN 978-7-5611-6520-1

 Ⅰ．①全… Ⅱ．①景… Ⅲ．①景观设计—作品集—世
界—现代 Ⅳ．①TU-856

 中国版本图书馆CIP数据核字(2011)第182901号

出版发行：大连理工大学出版社
   （地址：大连市软件园路80号 邮编：116023）
印 刷：利丰雅高印刷（深圳）有限公司
幅面尺寸：245mm×245mm
印 张：60
字 数：1300千字
出版时间：2011年9月第1版
印刷时间：2011年9月第1次印刷
策划编辑：苗慧珠
责任编辑：刘晓晶
责任校对：万莉立
版式设计：王 江 赵安康 张建实

ISBN 978-7-5611-6520-1
定 价：880.00元（全4册）

电 话：0411-84708842
传 真：0411-84701466
邮 购：0411-84708943
E-mail:dutp@dutp.cn
http://www.landscapedesign.net.cn

目录 **Contents**

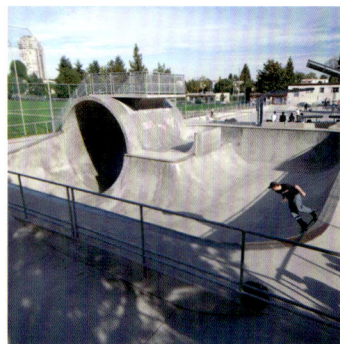

公园 _ **Park**

目录 Contents

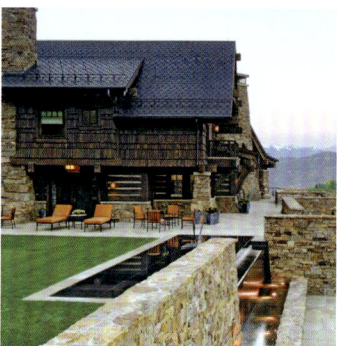

公园

index

修复亚马逊河生态景观并还之于民 —— 鹭之红树林公园

Restoring an Amazonian Landscape and Bringing It Close to People — Heron's Mangrove Park

撰文：Jimena Martignoni　　图片提供：João Ramid　　翻译：申为军

　　贝伦市位于巴西北部，地处亚马逊河入海口，一度退化的滨水景观经过改造后焕然一新，这说明城市边缘的自然景观也可以成为兼备休闲和教育功能的公共场所。鹭之红树林公园就是这样一个项目，它恢复了亚马逊热带雨林所特有的水生植物——海芋的自然生态形式，把这块被遗忘的土地改造成一个新兴的城市公园。该项目是贝伦最新完成的景观设计项目之一，现已成为当地重要的文化场所。

　　贝伦市为帕拉州的首府，亚马逊河的支流瓜玛河流经该市。如今，贝伦已经发展成一座典型的港口城市，经济和文化发展都高度依赖其所处的地理环境。自1616年葡萄牙殖民者在此建城后，贝伦历经沧桑，其中最明显的改变当属新千年伊始的几年间开展的大型城区改造计划。该计划由文化部长保罗·查韦斯主持实施，重点是改造包括建筑和滨水步道在内的历史城区，创建更多的公共空间和城市绿地。

1　鸟瞰图
2　鹭与小桥

总平面图

该场地原为海军所有，曾经是船坞的一部分，河岸边长满了成片的海芋；海芋生长极为茂盛，最高可达 3 m，但通常要将其修剪至 1 米左右，便于人们进出河岸并保持视线畅通。场地从北向南建有一堵 1.2 m 高的隔离墙，将这片植物完全隔离在场地外，以免受到任何形式的自然入侵。

2003 年，海军把这块 40 000 ㎡ 的土地交付给帕拉州，用于建造公共公园，由此拉开了改造工程的序幕。改造工作首先拆除了隔离墙、重新规划河岸，大片的海芋被保留下来，并任其自然生长。如今，那些海芋已经长至 3m 高，不但构成了场地的自然边界，也成为场地的重要组成部分，并且模糊了自然和人工设计之间的痕迹。海芋的外形有着十分显著的视觉效果，宛如大块的绿毯延伸至公园中。园路被设计成木质步道，引导人们在高大、茂盛的亚马逊河生态植物群落中穿行。这些蜿蜒的步道很受欢迎，有些高架步道通向河边，供人们俯瞰、眺望河流，有时候一转弯，水景又被茂密的植被所遮挡。

建造这个公园的首要目的是重建亚马逊河地区最具标志性的自然景观。在规划设计中，沼泽、森林、草地构成概念上的三大主题，并在设计中得以体现。在设计中，水的概念被明确地表达出来，它既是塑造公园各个景点的要素，也在文化上显示出亚马逊河重要的水生系统。此外，在贝伦极端燥热的气候中，水景能带来清爽的舒适体验。

在公园入口处，游客会首先看到地面上的一个形状不规则的喷泉，直接点出公园的水文理念。这个理念由著名景观设计师罗莎·克莱斯提出，并得到了专业工程师的技术支持。喷泉旁边是石阶跌水，注入到下方狭窄曲折的小溪中，小溪的两边种植了热带彩色灌木和花卉。这些植物镶嵌在小溪和高低错落的石刻边缘，从紫、橙、黄的各色凤梨科植物，到橄榄绿、褐色、铁红的多种狼尾草，色彩纷呈、千变万化。

小溪一直汇入中心湖，沿途经过几片开阔的草地，其间有几座小桥与公园的园路相连接。湖边聚集着大

群的草鹭、白鹭和野鸭，这些都是亚马逊河生态景观中常见的鸟类，公园的名字就由此得来。它们有时在岸边闲庭信步，当游客靠近时，又纷纷扎进水中。湖边设置了很多观鸟点，均为木质凉亭，上面爬满了藤蔓，旁边还设置了弧形坐椅供游客休憩，而低矮的石墙则方便了孩子们观察湖中的鸟类和其他生物。湖边种植了金鸟鹤蕉等姿态奇异的彩色植物；湖面上则种植着茂盛的水生植物，如王莲和水葫芦等。

除了这些浓厚的自然野趣，场地中还常常会看到有关城市历史的暗示。贝伦是由一个军港发展起来的城市，场地中随处可见这样的烙印，如湖心伫立的木雕船体结构、紧贴小桥栏杆的彩色龙骨以及被当做艺术品陈设在不同地点的独木舟和小船等。为完善场地的文化内涵，公园中还建造了一座海军博物馆，用来展示贝伦的历史背景。博物馆是半地下式的，位于公园中惟一的一幢建筑中，每逢周末都会吸引众多的游客。这幢建筑一层设有餐厅，在此不但可以享受巴西亚马逊地区的特色美食，还可以欣赏到河流美景和保存下来的植物群落——海芋。

这幢建筑采用木质结构，意在体现周边环境并融入其中。从餐厅露台延伸出一条100 m长的高架桥，一直通向岸边生长着海芋的河面上。桥体由木桩支撑，桥下是郁郁葱葱的热带植物群；桥的尽头是一个设有凉亭的观景露台，人们喜欢在炎热的下午来此纳凉，静静地欣赏流水，眺望远处的城市。

场地的教育功能也是要唤起当地民众的环境保护意识——为给蝴蝶和鸟类创造适宜的生存环境。设计师利用特殊的施工技术建造了两个大帐篷，里面可保持某一特定的温度。游客通过购买低价门票便可进入展区，在此既可了解聚居在这里的众多蝴蝶和鸟类，还能欣赏场地内部的水景设计。

1　象征文化传承的彩色龙骨
2　凉亭内部
3　被植物环绕的水景

The conversion of a degraded waterscape in the Amazonian city of Belem, in the North of Brazil and right at the mouth of the River Amazon, proves how some natural landscapes at the edges of cities can become responsible recreational and educational places for people. The Mangal das Garcas, or Heron's Mangrove Park, is a project that restored a natural formation of aningas—aquatic species endemic to tropical forests, and renovated an abandoned strip of land along the river as a new park for the city.

The city of Belem (State of Para) is located by the River Guama, a tributary of the Amazon, and consequently grown as a typical port city whose economy and culture highly depend on this geographical condition. Since its foundation in 1616 by Portuguese colonizers, the city experienced many changes, among which the most remarkable is a large urban renovation plan that has been implemented during the first years of the new millennium. This plan was put into action by the Secretary of Culture Paulo Chaves, and renovated historical areas, including buildings and the river promenade, and created new public spaces and green urban areas. This park is one of the latest landscape architecture projects built in Belem and one that has become an important cultural space for locals.

The site belonged to the navy and was part of a shipyard whose river bank had been profusely occupied by large groups of aningas (Montrichardia arborescens or mocou-mocou). These plants grow in dense clumps and can reach a maximum height of nine feet, the reason for which they had been cut to a height of 3 feet, at the most, in order to keep access and views open to the water. In addition, a 4 feet wall that crossed the entire site from North to South had been built to keep the aningal completely separated from the site and to avoid any natural invasions.

When in 2003 the navy authorities ceded 4 hectares of land to the State of Para for the construction of a public park, the first conversion works began. The wall was demolished and, after reshaping the river border, the aningal was left untouched to allow its natural growth. Today the aningas are 3 meters tall and more; they not only constitute the natural edge of the site but they are a part of it, making the limits between nature and manmade design more flexible. This green mass seems to enter the park because of its striking visual appearance, and the park's paths, shaped as wooden walkways, pass through the tall, thick, Amazonian plant community. People love walking these winding pathways; certain spots of these floating wooden boardwalks overlook the river and after some unexpected turns the sight of the water becomes screened by the plants.

The primary objective for this park was to re-create the most emblematic landscapes of the Amazonian region. Swamps, forests and meadows are the three themes that conceptually configured the layout, and which were built in the site. In a very explicit gesture, water was chosen as the main visual and aesthetical connecting element throughout the park, and as a cultural reference to the very significant aquatic system of the Amazon. In addition, the refreshing presence of water turns the extremely hot weather of Belem into a much better experience.

At the entrance, an irregularly-shaped water fountain built at ground level becomes a first feature which introduces the visitor to the hydrological concept that was developed in the park by landscape architect Rosa Kliass, with the technical assistance of specialized engineers. Nearby, a cascade that emerges from a terraced stone structure falls down to a narrow meandering stream whose borders are planted with tropical colorful shrubs and flowers. The plants edges the stream and the different levels of the stone composition, offering a fantastic palette of colors that range from the purple, orange and yellow hues of the Aechmea fulgens (coralberry) and other air pines and bromeliads, to the olive green, brown and oxide hues of the Pennisetum.

Along its flowing journey into a central lake, this water

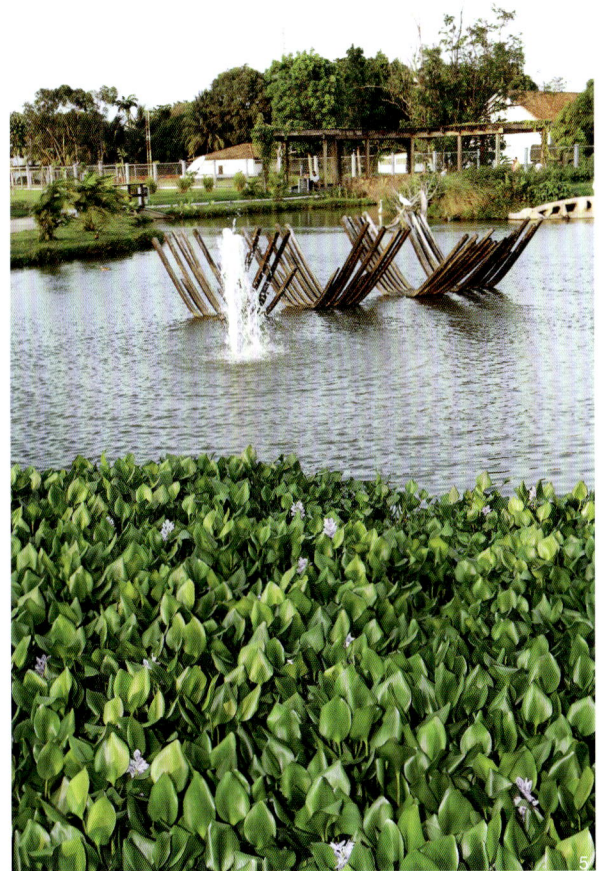

1、2 凉棚

3 眺望台

4 中心湖与小桥

5 水景

course goes through some green open meadows and is crossed by a couple of small bridges that connect to the park's path system. The lake edges are crowded with purple herons, white herons and wild ducks; these birds, native to the aquatic landscapes of the Amazon and which actually gave the park its name, walk placidly along the lake borders or dive into the water while kids and adults gather together to watch them. Around the lake, many spaces were designed as bird-watching spots: a wooden pergola covered with vines and a curvilinear bench where couples sit around for hours, and low rocky walls on which kids lean or sit to gaze at the aquatic fauna. Colorful plants with exotic forms, such as Heliconia psitacorum (Parrots Beak Heliconia) define the lake's edge and thick masses of aquatic plant species, such as Victoria regia (Amazon lily) and Eichornea crassipes (water hyacinth) cover the water surface.

Besides the strong experience of nature that the site offers, it constantly makes a clear reference to the local history of the city-port of Belem. Allusions to the city's naval history can be found all over, such as a sculptural wooden ship structure in the center of the lake, colorful keels attached to the bridges handrails, and canoes and small boats located at different spots of the site exhibited as art pieces. To complete the series of elements that refer to the site's cultural significance, a Navy Museum exhibits the nautical history of Belem; half buried and part of the only building in the park, this place attracts large groups of locals and visitors every weekend. On the first floor, a restaurant offers typical dishes of the Amazonian region of Brazil and provides amazing views of the river and the restored aningal.

The wooden structure of this building seeks to reflect the surrounding environment and to integrate with it. A 100 meter-long elevated pier, which stands over wooden pilings emerging from the plant composition, starts at the restaurant's terraces and continues into the river, framed by the aningas. At the very end of the pier, a gazebo-like structure becomes a favorite spot for people during the hot afternoons; people stand and relax while gazing at the water and the distant historical city market of Belem.

The educational purposes of the site are also related to the environmental awareness of the city's residents. In order to recreate the natural habitats of butterflies and birds, the project includes two pavilions which were built with special construction techniques and in response to specific temperature requirements. Visitors have to pay a low entrance fee to get into these exhibiting areas; they can learn about the many native species that live here together while enjoying the water features designed inside these structures.

混凝土的构筑 —— Metro滑板公园

Avant Concrete — Metro Skate Park

撰文：Derek DeLand 　　翻译：王睿瑾 孙路

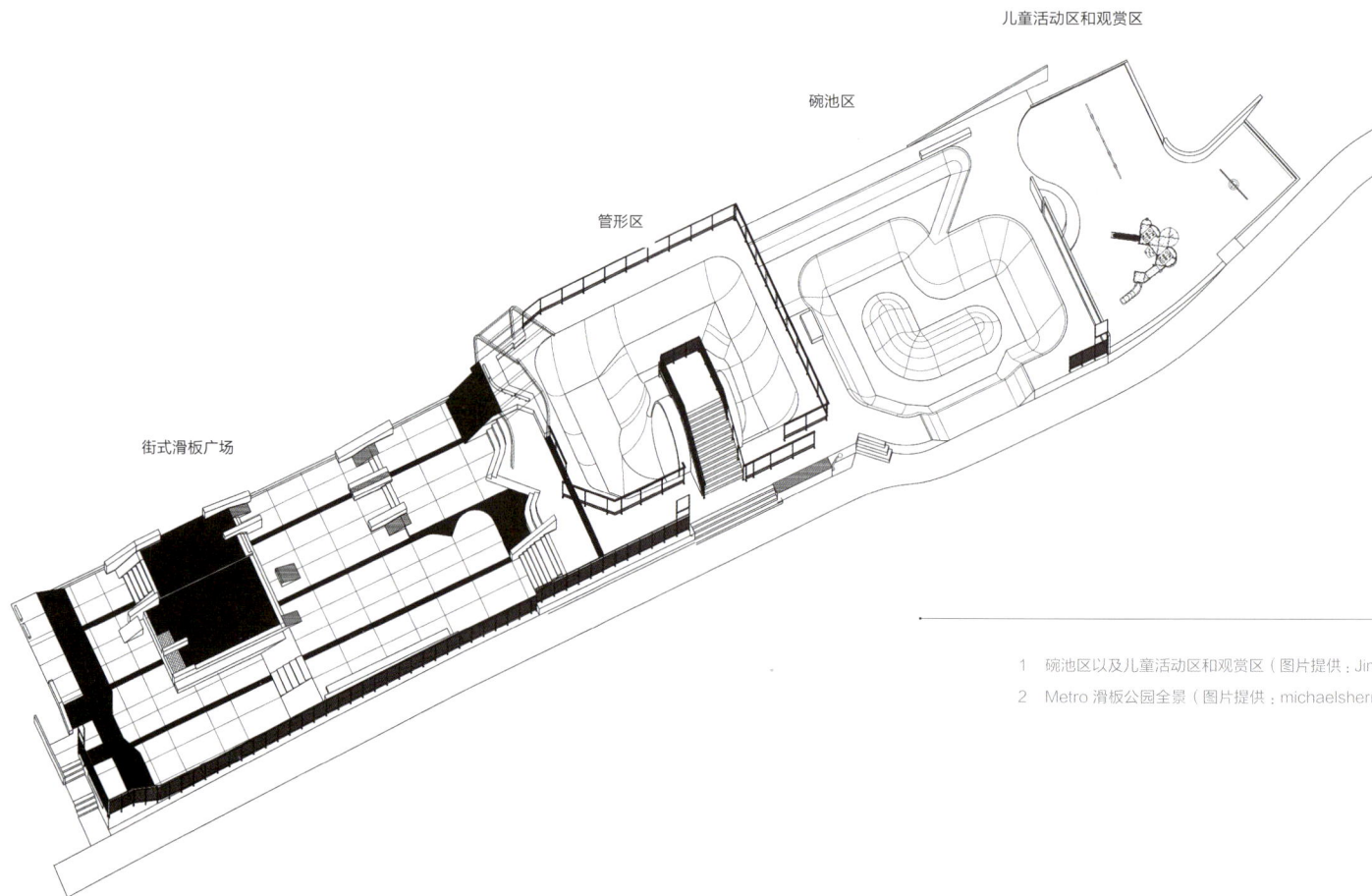

儿童活动区和观赏区

碗池区

管形区

街式滑板广场

1 碗池区以及儿童活动区和观赏区（图片提供：Jim Barnum）
2 Metro 滑板公园全景（图片提供：michaelsherman.ca）

近年来，滑板运动为景观设计的发展带来了新的设计灵感，滑板公园的出现不仅反映出人们对这项体育运动的热爱，更体现出滑板运动爱好者试图通过建筑美学来表达自我的愿望。该项目证明了滑板公园景观规划的发展已步入新的阶段。

滑板公园可以多角度地展示现浇混凝土的可塑潜力，并很好地诠释这种普通原材料的自然美。设计师对混凝土的运用就像大师弗兰克·盖瑞对钛的运用——别出心裁而又恰到好处。滑板爱好者对混凝土的喜爱并非只是因为它的可塑性，而是因为它的本质。因此，一个真正完美的滑板公园需要在深度和广度上充分体现出设计师对混凝土的自如运用以及对其本质的表达。

广泛参与的设计

该项目成功的关键在于设计师在整个设计过程中反复征集了不同类型滑板运动爱好者的意见，包括概念设计、整体方案调整、方案优化。根据征集的结果，设计师建造了滑板爱好者情有独钟的造型——管形区。但儿童更喜欢流线型的设计，而青少年则钟爱棱角鲜明的设计，所以设计师又将儿童活动区和观赏区与碗池区连接在一起。设计方案公布后，赢得了良好的反响。

四个区域的完美融合

该项目建在高架路与棒球场之间的一条狭长区域内，包括四个功能区，即专供街式滑行的广场、用于练习过渡穿越的管形区、作为高手驰骋天地的碗池区以及儿童活动区和观赏区。四个功能区连成一线，其整体的连贯性给人以大气且柔美的感觉，将不同的滑板形式和规则通过建筑形式完整地呈现出来。

1.街式滑板广场

该区域包括了即街区和圆石沟区，整体呈矩形，体现出一种都市广场的气息。由于街式滑行是滑板运

动中最基础且最流行的技巧，所以这一区域的面积占整个公园面积的 60%，设计上也刻意结合了滑板的创新和技巧。

即街区独特的造型使在这里进行街式滑行的速度更快，高超的技巧展示使滑板运动几乎变成了武术表演。即街区又分为主滑道区和小型滑板区，分别由深灰色混凝土滑道与小型齿状滑道组成，为滑行带来灵动的节奏感，滑板爱好者们可以自由选择花式技巧，时而垂直、时而平行。而且这些形状迥异的滑道使整个区域错落有致，又自然地划分出了不同的滑板技巧区。

圆石沟区是滑板公园中的独特元素，也是滑板爱好者挑战高难度技巧的训练地和摄影爱好者拍摄时的独特视角。设计师在凹凸不平的地面堆砌了三层坐椅，正如弗兰克·劳埃德·赖特大师设计的"流水别墅"一样，进行了一番艺术与自然的对话。

与传统滑板场地上有钢质边缘的光滑混凝土构筑物不同，该项目使用了多种质地与花色的材料。设计师用花岗岩作横挡，将镀锌钢镶嵌在某些构筑物的表面，龟裂状的混凝土使墙面看上去更有厚重感，框边产生的阴影衔接了各个不同区域。滑板爱好者们在滑行过程中忽上忽下，宛如一个个跳跃的音符。

2. 管形区

管形区是滑板公园最独特的标志性区域，不仅造型独特，而且还设有一对弧形的钢质遮阳棚。该区域的坐椅和看台宽大舒适，加上弧形遮阳棚，为人们提供了一个很好的休息交流区域。

目前，许多滑板公园的规划只关注了场地的变化，而该项目则别出心裁，将地面构筑物设计得错落有致。管形区的主体部分是有顶盖的圆柱形构筑物，设计师称之为"Bonsor 管形结构"，属于景观与建筑相融合的设计，也是公园中与众不同的标志。这种设计元素会使人联想起英国女性建筑大师扎哈·哈迪德或荷兰建筑大师雷姆·库哈斯的作品。管道内部直径为 5.48m，与毗邻的高架路相协调，而且 2.74m 的内半径正适合滑板运动。管道的北面延伸出一个平台，与碗池区相连，使两个区域紧密地联系在一起。

弧形遮阳棚将即街区与管形区衔接起来，除了可以为人们遮风挡雨外，它还与管形区以及毗邻的高架路遥相呼应。弧度造型打破了滑板公园的整体平面感，使整个公园的立体感十足，也使钢质材料与混凝土的结合相得益彰。混凝土长椅的设计灵感来自于工业装卸码头，靠背的设计使其成为小憩的最佳选择，但是精彩而刺激的表演绝对不会使您昏昏欲睡。

3. 碗池区

该区域因其外形酷似大碗而得名，这里适于初学者使用，难度稍低，也是应儿童滑板爱好者的要求而设计的场地。碗池区充分展示了混凝土材料塑造出的流线型的平面特点。

Metro 滑板公园的碗池区是加拿大第一个采用定制弧形模块再进行衔接的案例。定制的模块按照一定的高度、半径和弧度用 EcoSmart 混凝土、HardCem 硬化剂以及微纤维现场混合浇筑而成。设计师在每道衔接

1　弧形遮阳棚（图片提供：Peter Whitley）
2　涂鸦拼图（图片提供：Jeff Culter）
3　充满童趣的瓷砖墙（图片提供：Jeff Culter）
4　街式滑板广场（图片提供：Jeff Culter）
5　弧形遮阳棚为人们提供了休息的空间
　　（图片提供：Jeff Culter）

处都涂了一层光亮的深色漆，以减少摩擦并勾画出了碗池区边缘的轮廓。由于碗池区的边缘没有设置顶盖，因此滑板爱好者在这里练习时无需担心会被顶盖钩到，而且这里的每个角都是平滑的，可以使滑板动作更流畅完整。

碗池区外壁镶嵌的图案是由当地的涂鸦艺术家们设计的，再由当地的孩子们拼贴完成。这些图案充分表达了艺术家们的想法，人们能从中感受到他们的自豪感。与此同时，这种设计也有效地减少了社区中其他区域的涂鸦现象。

4. 儿童活动区和观赏区

这是一个色彩缤纷的区域，配有相应的游乐设施和塑胶保护层，仅在入口楼梯、区域边界以及瓷砖墙等处使用了混凝土。与活动区相对的是一面颜色亮丽的瓷砖墙，墙上的图案也是由当地孩子们绘出的，仿佛这里是只属于他们的天地。设计师将活动区与初学者练习区并置，并使其同时处于看台的视野中，孩子们可以在这里伴着滑板度过快乐的童年。

从某种意义上来说，Metro 滑板公园体现了滑板运动由初级向高级发展的过程——从初级的嬉戏玩耍到简单的技巧练习，再到流畅的穿越与现代街式滑板，整个过程都能在滑板公园内完成。公园的建造全部采用了环保的粉煤灰混凝土，体现出伯纳比市对城市以及年轻人未来的重视。

In the past few years, skateparks have been pushing landscape architecture in innovative new directions. The emergence of the innovative skatepark typology reflects more than enthusiasm for a sport, it also represents skateboarders' desire to define themselves by their own authentic architectural aesthetic. Metro Skate Park is a milestone of the skatepark's arrival as the legitimate landscape architecture of skateboard culture.

Skateparks showcase sitecast concrete's creative potential, unlocking its beauty via a multifaceted exploration of the raw material's polymorphous sculptural nature. They see concrete the way Frank Gehry sees titanium. Skateboarders love concrete not for what it can imitate, but for what it truly is. A really good skatepark is an expression of the range and depth of what concrete can be if it behaves precisely like itself.

Democratic Design

A key reason for Metro's excellence is its extraordinary depth of skater consultation during the design process. Instead of one design consultation, different categories of skaters were consulted repeatedly by designers through the process. Adolescent skaters were asked at one meeting for conceptual sketches, at a second meeting for overall schemes and social areas, and at a third meeting to refine the chosen concept. These workshops showed clear desire for an iconic element, which lead to the inhabited fullpipe. Separate elementary school consultation generated fluid forms in contrast to the adolescents' preferred angles, and linked playground with flow bowl. Completed design concepts were publicly displayed, with positive response.

Four Part Harmony

Built on a long narrow urban margin between an aerial commuter train and grass baseball diamonds, Metro arrays four concrete personalities in a row, creating a sequential whole. Metro is distinguished by its ambition, aesthetic range, fine detail, and architectural articulation of skateboarding's different disciplines. Metro Square plaza is for Street skating, Bonsor Pipeline for Transition skating, the Dogleg bowl accommodates Flow skating, and the Playground lets young children safely watch.

1. Metro Square Street Plaza

This section is a crisply crystalline matrix of orthogonal evoking a big-city urban plaza. Street is skateboarding's most popular form, so this represents 60% of the park's area. The creativity and sophistication of detailing has been pushed hard, limited only by the function of small wheels.

The geometric architecture of the plaza evokes the straight lines and quick movements of street skating, more martial art than sidewalk surfing. The composition makes sharp distinctions between horizontal and vertical, leaving

it to skaters' athleticism to connect elements. The plaza's ordering system consists of a major grid of dark gray concrete bands and a minor grid of knifelike sawcuts that create a mathematical sonic rhythm when ridden. These contrasting bands and cuts lend visual variety to the plaza, in some cases carrying through elements as they change elevations and angles.

Three seating stairs poured around a site-excavated boulder, the Boulder Gap is the plaza's signature element. It creates a dialogue between Natural and Artificial, similar to F.L. Wright's Fallingwater. The gap challenges skaters, spices up photographs, and gives local identity. When magazine photos show a professional skater kickflipping the gap, Burnaby kids are proud that it's their park.

Traditionally, skateparks were smooth concrete volumes with steel edging. In contrast, Metro embraces multiple materials and textures. Blocks of granite are mounted as ledges, and sheets of galvanized steel are embedded flush as a surface material in some elements. Split-face concrete adds visual depth to a wall. Deep reveals create shadowlines that articulate junctions between elements, making objects visually pop off of the plaza surface as sharply as the skaters who use them.

2. Bonsor Pipeline

The Pipeline section is the most dramatic of Metro's sections, featuring an imposing concrete fullpipe, and twin angled steel sunshades. Metro's social strategy is to draw passive observers into the active experience; the fullpipe's grand stair and viewing platform excites visitors with an adventure, whereas the protected sunshade seating welcomes them to a relaxing social space.

Where most skateboard parks confine themselves to variations of the ground plane, Metro Skate Park has higher aspirations, turning a surface-oriented architecture upside down. This fullpipe structure is named the "Bonsor Pipeline". This is landscape verging on architecture, creating what is essentially a cylindrical outdoor room with a roof deck. The structure recalls elements of projects created by Zaha Hadid or Rem Koolhaas. The fullpipe is Metro's signature urban-scale element. The pipe's internal diameter of 18 feet is massive enough to respond to the scale of the neighboring Skytrain track, yet has a skateboard-friendly internal 9' radius. Approached from the north, it evokes a futuristic ceremonial platform. The fullpipe, its adjacent hip, and the tight transitions on the bowl's central deck are all massaged together in carefully considered junctions of impressive craftsmanship.

The sunshade roofs between plaza and transition sections fend off sun and rain, but also serve an architectural function. Their shapes act as an urban-scale counterpoint to the fullpipe and the Skytrain, breaking the horizontal plane of the park and creating spatial interest. The steel creates welcome material contrast to the dominant concrete, and

1　从儿童活动区和观赏区看滑板表演
（图片提供：Jeff Culter）
2　街式滑板广场细部
（图片提供：Jeff Culter）

delineates the space over the social seating area. The generous concrete benches feature pipe backrests inspired by industrial loading docks; these backrests can also be used as short-term seating but discourage unwanted sleeping.

3. Dogleg Bowl

This section expresses concrete's ability to create an uninterrupted, fluid surface. Intended for beginners and named for its shape, the Dogleg bowl was designed to respond to the wishes of elementary school-aged kids in the consultation group.

The mosaic on the Dogleg's wall was drawn as a graphic by local graffiti artists during the consultation workshops, and then installed by local kids during construction. The use of graphics created by community youth allows them to express themselves and creates a sense of ownership, reducing graffiti in the area.

4. Playground Play Area for Observation

The children's playground features colourful playground equipment and child-friendly bark mulch, with concrete just used for access stairs, borders, and a backing wall

for tiles. Facing the playground is a tiled wall grid of brightly-coloured ceramic tiles, set off by a reveal. The tiles were painted by local children and made the space feel like theirs. The playground is intentionally juxtaposed with the beginner area of the skatepark, with concrete viewing stairs allowing visual access. This encourages growing playground kids to keep playing by picking up a skateboard.

In a sense, the whole project can be read as a developmental timeline, tracing skateboarding's evolution from child's play, to simple carving, to transition skating, to modern street skating. Built entirely with ecologically friendly high-fly-ash concrete, Metro speaks well of Burnaby's respect for its youth and their future.

1　体现了滑板公园设计的广泛参与性
　（图片提供：Jim Barnum）
2　阳光中的"Bonsor 管形结构"
　（图片提供：Jonathan Mentzos）

乌龟徜徉在小岛上 —— 富兰克林儿童公园

Where a Turtle Walks in the Island — Franklin Children's Garden

撰文：Pedro F Marcelino　　　图片提供：Janet Rosenberg + Associates　　　翻译：张璐

从多伦多市中心到湖心岛，坐船仅需短短 10 分钟。富兰克林儿童公园就坐落在这片令人耳目一新的自然区域的中心地带，公园的建设灵感来自于受欢迎的儿童丛书《小乌龟富兰克林》。整个公园是开放式设计，没有围栏，由若干各具特色的公园互相连接而成。各个年龄段的孩子们可以在这里通过畅游小乌龟富兰克林的自然世界来学习和了解湖心岛生态系统的

构造，在学习的过程中孩子们可以探索、发现并相互交流。

儿童公园的建设把教育孩子保护环境摆在首位，以此为目的的景观建筑在多伦多市屈指可数。居住在高楼林立的大城市中的孩子们接触自然、与自然互动和领略自然的机会越来越少，而跟大自然亲密接触才能让孩子们真正学习到并切身体会到保护自然环境的

重要性，感受到自然环境的细微变化以及欣赏大自然的美丽景色。在设计建造这座儿童公园时，Janet Rosenberg+Associates 的设计团队想方设法使整个自然环境富有趣味性、让孩子们喜欢这里并且流连忘返。众所周知，孩子们通常从书籍和电视中获取知识，而小乌龟富兰克林的自然世界就是孩子们最好的选择，他们既能在书上读到这只小乌龟的故事，还能在电视

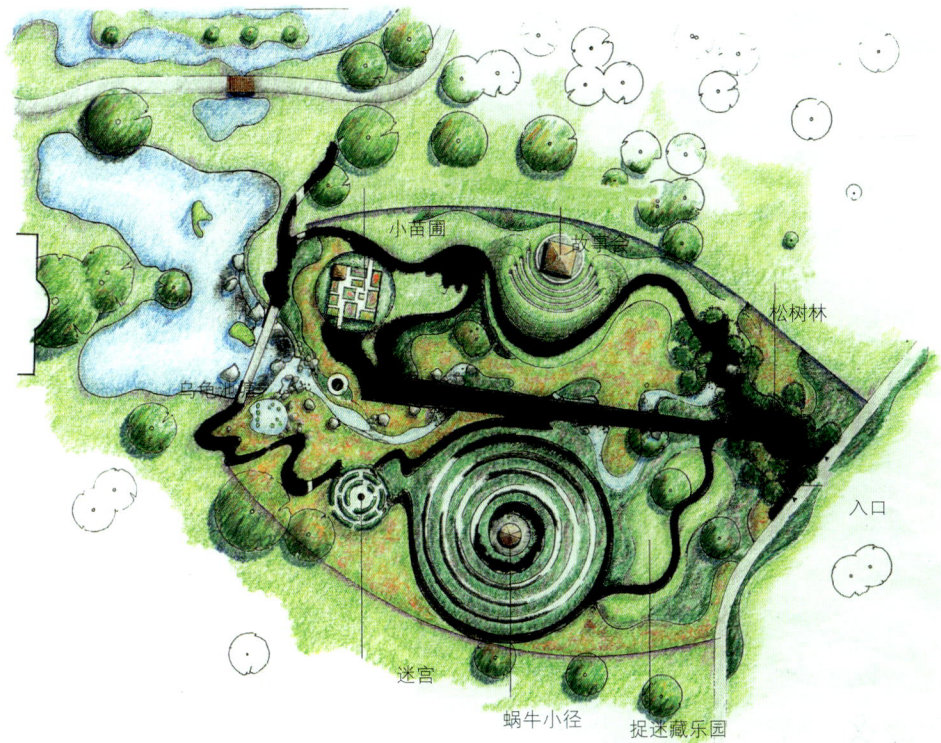

公园平面图

（平面图标注）
小苗圃
故事会
松树林
入口
迷宫
蜗牛小径
捉迷藏乐园

1 孩子们乐于接触故事中的角色的雕塑
2 园区指示牌

上看到同名动画。孩子们也会因为看到了书上描述的小乌龟富兰克林，而乐于跟公园里生活在湿地中的乌龟近距离接触。在这片天然湿地里还居住着很多不同的生物，如青蛙、小鱼和小鸟等，孩子们还能学习到水是如何通过植物进行过滤的。

景观设计从草坪和树木入手，将这块面积约为16 000平方米的场地转变成若干相互连接的花园：有松树林、蜗牛小径、捉迷藏乐园、乌龟池塘、小苗圃以及故事会。公园里有一个露天剧场，经常会有人在这里给孩子们讲故事，当然听故事是免费的；还有一间无障碍树屋，方便轮椅通行；另外，还有葡萄藤搭起的小隧道、生活着许多乌龟和青蛙的天然湿地和垂直水幕；还有一组（共7件）铜质雕塑，是小乌龟富兰克林和他的伙伴们。孩子们平时总在铜像上爬来爬去，衣服把铜像擦得铮亮。游客们沿着公园内的主路前行，穿过公园中心，便大致可以领略到公园的全貌。主路向外延伸出许多蜿蜒的小径，好像在催促游人赶快投入到寻宝的旅程中。

设计遇到的难题是如何才能借助小乌龟富兰克林这个角色突显自然的重要性，而不是建成千篇一律的主题公园。于是，设计师与《小乌龟富兰克林》一书的作者与插图者进行了密切合作。要想成功建设开放式公园并能让孩子们与大自然亲密接触，作者的作用是不可小觑的。该项目所在地的自然生态环境告诉人们，由于不断开发多伦多市滨水区域，当地乌龟的栖息地明显减少。设计师修建新的地形来种植各种各样的植物，创造出仙境般的玩耍嬉戏之地，例如蜗牛小径——那是一条呈螺旋式上升并直通山顶的小路，两边排列着密密麻麻的灌木柳。孩子们走在小径上一边兴奋地尖叫着，一边又不时地躲进灌木丛中顽皮地露出头来。在小苗圃里，孩子们可以更主动地参与活动，给植物浇水、观察它们的生长情况，还为植物修剪枝叶。对于那些生活在城市中没有花园可以照料的孩子们来说，这是一个接触自然的绝好机会。另外，孩子们还可以站在由 7 个水罐组成的水景下任凭水流自上流下，体会与植物一样"被浇灌的感觉"。在酷热且潮湿的夏天，孩子们玩儿得兴致盎然。

电影《梦幻之地》里描绘的情景变成了现实。只要用心去营造，大自然就能重新回到人们身边，富兰克林儿童公园就是这样建造起来的。儿童公园初见形状，现在已经成为多伦多市滨水地区的一处标志景观。借助书中人物和公共艺术，公园给孩子们提供了发挥想像和了解大自然的机会。富兰克林儿童公园不仅是一项了不起的设计，而且也令家长们感到十分满意。

Toronto Island is a short 10-minute ferry crossing from Downtown Toronto. Franklin Children's Garden is located in the heart of this unexpected natural area. Inspired by the popular Franklin the Turtle books, the park is an interactive barrier-free environment consisting of various interconnected gardens where children of all ages can learn about the Toronto Island's ecosystem through a playful realization of Franklin's natural world, which they are free to explore and interact with.

The garden is one of very few landscape architectural projects in the city of Toronto that places the environmental education of children at the forefront. Children living in dense urban settings such as Toronto have increasingly little contact, interaction and engagement with nature on a level where they can learn and experience first-hand its importance, its subtle changeability, and its beauty. In the design of the park, Janet Rosenberg + Associates aimed at making nature fun, exciting, almost magical. Drawing upon a concept that children know most from reading and television, Franklin the Turtle's natural world, seemed like

1 盛夏季节的故事会

2 书中角色的雕塑

3 孩子们为植物浇水

the best option to achieve these results. Kids are naturally presented with real turtles in the park, to ensure there is a meaningful connection between the literary turtle and the actual wetland ecosystem. The wetland is also home to different species of frogs, fish, and birds, and the perfect setting to learn about how water is filtered through the native plant species.

Starting from a basic landscape of grass and trees, this four-acre site was transformed into a series of interconnected gardens: Pine Grove, Snail Trail, Hide & Seek Garden, Turtle Pond, Little Sprouts Garden, and Storybook Place. The park includes an amphitheatre, which is used for free storytelling programs, a wheel-chair accessible tree house and vine tunnel, a natural wetland that is home to different species of indigenous turtles and frogs, a custom water feature, and public art with seven child-accessible bronze sculptures of Franklin the turtle and his storybook friends. By now, the sculptures have a gleaming bronze patina from the buffing provided by thousands of clambering children. A main pathway leads visitors through the centre of the park and sets up its geometry. Several smaller meandering trails stream out of the central path, urging visitors to embark on an exploration of its whimsical treasures.

One of the project's challenges was how to avoid a theme park approach but instead to use the Franklin character as a springboard for a celebration of nature itself. The designer worked closely with the books' author and illustrator, who were instrumental in realizing a vision for a barrier-free public park where children can incorporate

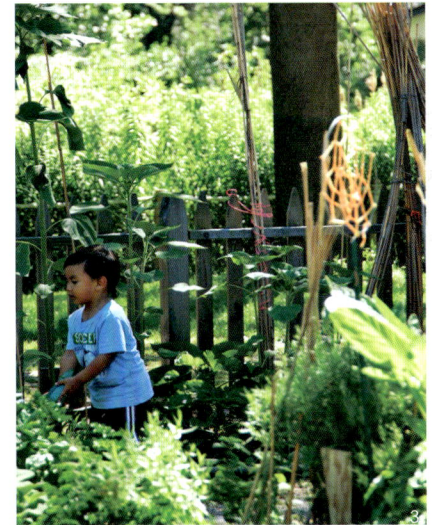

nature in their play seamlessly. The site's natural ecosystem was an ideal backdrop to highlight the shrinking habitat for indigenous turtles in the face of expanding development on the Toronto waterfront. A new topography was also developed to incorporate diverse plant materials and create magical play areas such as the Snail Trail—a spiraling path going uphill, lined densely with willow shrubs. As children move through the path, they literally squeal with delight as they disappear with only their heads poking out from the vegetation. In the Little Sprouts Garden children

can take on a more proactive role and water the plants, see them grow over time, and become engaged gardeners. This is especially great for kids who live in the city and don't have a garden of their own to tend to, the proximity of Downtown making it accessible. On the other hand, children can walk over to the water feature made up of several watering cans and experience the "feeling of being a plant" themselves, as they stand under the cans and feel the water flow over them. A crowd pleaser in Toronto's torrid (and humid) summer days.

The message from Field of Dreams turned out to be true. If you build it—or at least if you build it right—nature will come. And so it has at the Franklin Children's Garden. Slowly the garden gained shape and is now a mature element of the Toronto waterfront. This garden provides children with opportunities to develop their imaginations and expand their understanding of the natural world by using a literary figure and public art as a springboard. Not only is it great design, it also makes parents smile.

敏感生态环境中的可持续设计 —— 温特沃斯公共区游乐场

Sustainable Design in Sensitive Habitats — Wentworth Common Playground

撰文 / 图片提供：HASSELL 设计集团　　翻译：牟誉

该项目由澳大利亚 HASSELL 设计集团负责设计，是悉尼霍姆布什湾悉尼奥林匹克公园中最新完成的区域。

悉尼奥林匹克公园占地近 500 万平方米，是悉尼最大的城市公园之一，包括复原的自然景观和敏感的生态环境区，如林地、盐沼和潮间带湿地等。1997 年，HASSELL 设计集团联合彼得·沃克合伙人景观设计事务所、Bruce Mackenzie 等设计公司共同提出了公园的规划方案。该方案成为在土壤退化的城市中建设现代公园的范例，也为 21 世纪甚至更遥远的未来提供了城市公园建造的参考。

悉尼奥林匹克公园是特别为 2000 年举办悉尼奥运会和残奥会而建造的。虽然奥运盛事又传递到了其他城市，但是该公园因其特殊的自然环境和先进的体育设施而成为了悉尼市民生活中不可或缺的一部分。公园将建筑与有效管理和教育活动结合起来，其中教育内容侧重于挽救和保护自然环境、娱乐、文化、艺术、历史和科学。园区为奥运会提供的主要比赛场馆均设在霍姆布什湾，以"绿色奥运"为目标、坚持可持续发展是对奥运会所有工程的要求，这一要求也将被今后的奥运会举办城市继续遵循下去。

奥运会以后，公园又有了进一步的发展，不仅成为活跃的社区互动中心，还成为了吸引着越来越多的澳大利亚人和国际游客的景点。每年有 850 多万人在此参与种类繁多的休闲、娱乐、文化和教育活动。悉尼奥林匹克公园已被选为悉尼最受欢迎的公园之一，甚至被称做是西悉尼人的"后院"。

温特沃斯公共区是公园里 10 个独立的娱乐区之一，位于霍姆布什湾的标志性区域之中，从金字塔形的建筑上可以观赏到湿地、红树林、霍姆布什湾以及通向罗德斯的巴拉玛特河。经过设计师的巧妙设计，这些建筑能够经受过度的破坏，并为游客指引方向。温特沃斯公共区就是为社区聚会、烧烤、野餐、娱乐、进行体育运动和马术活动而设计的。

这块面积为 90 000 ㎡ 的土地曾经是国家砖厂，现如今则成为了具有重要价值的重建景观，是金绿色铃蛙（Green and Golden Bell Frog）和其他蛙类以及几种淡水鸟类的主要栖息地。

景观设计团队所面临的挑战是把部分地区改造成能引发人们好奇心和冒险精神的游乐场。为此，设计师特别将游乐设施巧妙地安置在原有的白千层灌木林中。由附近的水质处理池、Haslams 河提供水源的一系列浅池和水渠体现了游乐场中水的主题。

无论在功能还是美观方面，游乐场的设计都要考虑到极为敏感的生态环境。作为深受欢迎的冒险游戏区，这里吸引着那些很难有机会接近自然的悉尼家庭前来探索灌木林。它缓解了公园其他地方的原有游乐

1 娱乐和商业　　　12 二百年公园南部　　23 北部水景
2 商业和住宅　　　13 山路停车场　　　　24 废弃物处理厂
3 澳大利亚中心　　14 休闲设施　　　　　25 奥运村
4 草地　　　　　　15 临时狂欢节　　　　26 Newington 房屋和教堂
5 红树木和湿地　　16 广场　　　　　　　27 巴拉玛特河
6 射箭场　　　　　17 主要的运动和娱乐建筑群　28 千年标志物
7 娱乐区　　　　　18 淡水湿地　　　　　29 渡口
8 砖窑　　　　　　19 盐沼湿地　　　　　30 奥林匹克公园火车站
9 Newington 树林　20 Haslams 河　　　　31 千年公园入口标志物
10 高尔夫练习场　　21 北 Newington 草地　32 棚架
11 网球中心　　　　22 遗产建筑　　　　　33 码头

scale 1:5000 at A0
job no. 2664
dwg. no. MPLA084C
date 27.06.97

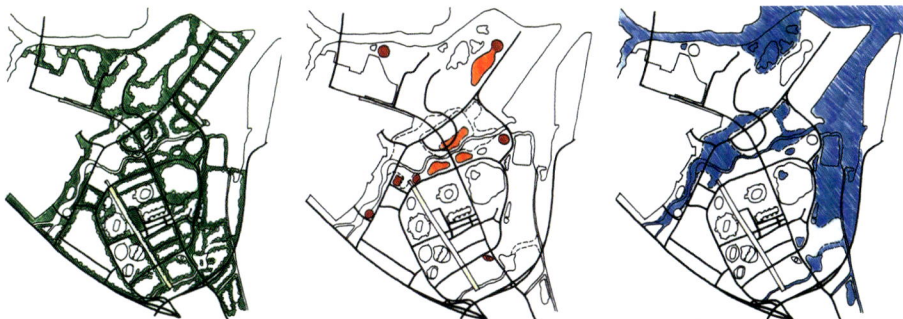

设施被过度使用的情况，也为稍大的孩子们提供了更多具有挑战性的游戏体验，并带给年纪稍小的孩子们想像的空间。

有几条小路引导孩子们进出游乐场。游乐场中有一些游乐设施是特别设计的，雕刻精美的木质"盒子"，与周围的木麻黄乔木林和白千层灌木林组成的广阔的公园环境相呼应。这些设施模拟了从森林到公园的道路，其中包括隐蔽的鸟类栖息地、穿越森林的隧道、一个开阔的日光休息室和一座开放的户外平台。设施的特殊尺度也有助于丰富孩子们的游戏经验，比如一些方形器具就像是超大型的玩具。游乐场的规模、不同器械的多功能特质和与自然相整合的特点都使温特沃斯公共区游乐场有别于其他的游乐场所。

天然和可回收材料的利用和抛光（如采用与四周石楠相同的样式和纹理）清晰地传达了设计师追求自然的设计理念。沿着灌木林和树林里延伸的小径，这些巨大的盒子也将游乐场"游戏旅程"的主题延伸到一系列户外空间中。

大型攀爬设施放置在沙地上，可动的雕塑悬于半空中。这个由设计师 Louise Pearson 设计的艺术作品名为"飞鸟"，与游乐场的自然环境相一致。其创作灵感来源于栖息在这片湿地中的鸟类飞行时的壮观情景，设计师还设计了几只由黄铜制成的鸟，这些"小鸟"站在线上可以随风摇摆，使整个雕塑更具有表现力。这件艺术作品完整生动，与整个公园的设计理念十分吻合。

游乐场变化的尺度、光影和围合感吸引着孩子们，而且他们每次都能获得新的体验。公园内尽量使用天然材料以加强与环境的和谐。为遵循可持续性原则，公园内还使用了回收再利用的木材和沙岩，其纹理与灌木林和天然草地的景观形成明暗对比。

温特沃斯公共区是一处重要的娱乐休闲场地，也是 2000 年悉尼奥运会遗产的重要组成部分。公园顾问委员会主席 Penelope Figgis AO 说："这个公园体现了当今社会价值的变化。该场地是上几代人无视环境、滥用土地的典型，也反映了全世界所面临的环境问题。悉尼奥运会和奥运会公园是我们对待土地的态度发生变化的象征——必须保护、恢复、治愈这片土地，使之更具有价值。该项目证明了没有什么破坏是无法修复的。"

Designed by HASSELL, Wentworth Common Regional Playground is the latest precinct of the Sydney Olympic Park at Homebush in Sydney's inner west to be completed.

The Sydney Olympic Park comprises nearly 500 ha of restored landscapes and sensitive habitats including woodland, salt marsh and intertidal wetland and is one of the largest metropolitan parks in Sydney. The concept plan for the parklands was prepared in 1997 by HASSELL in association with Peter Walker and Partners and Bruce Mackenzie Design and has become a benchmark for modern parks established on degraded lands in urban areas, providing a world model for metropolitan parklands of the 21st century and beyond.

The 2000 Sydney Olympic and Paralympic Games were the catalyst for the creation of Sydney Olympic Park. Although the games have moved on, the legacy of the park, with its extraordinary natural environment and state-of-the art sporting and event facilities is now integral to the life of Sydney residents. The parklands combine physical elements with an expandable and ambitious management and education programme focused on remediation, conservation, recreation, culture, the arts, history and science. The parklands provided the setting for the major Olympic Games venues at Homebush Bay. Known as the "Green Games", sustainability was a key requirement of all Olympic projects and for those which have followed.

Since the Games, the parklands have further evolved and are now a place of lively community interaction and a must-see attraction for increasing numbers of Australian and international visitors.

Each year more than 8.5 million visitors enjoy its diverse range of leisure, entertainment, cultural and educational activities. It has been voted Sydney's most loved park and is considered to be the 'backyard' for the people of western Sydney.

Wentworth Common is one of 10 individual recreation precincts within the parklands and is situated near one of the prominent Homebush Bay Markers; these are ziggurats with views over the wetlands, mangroves, Homebush Bay and the Parramatta River to Rhodes. They were invented by the designers as novel ways to accommodate excess spoil and provide orientation points for visitors to the parklands. The Wentworth Common precinct was designed for community gatherings, barbeques, picnics, recreation, sports and equestrian activities.

The 9ha precinct, previously part of the State Brickworks site, is now a reconstructed landscape with significant value as a primary habitat of the Green and Golden Bell Frog and other frog species, as well as several types of fresh water birds.

The challenge for the HASSELL landscape architecture team was to transform part of the precinct into a new regional adventure playground, one which invites curiosity, introduces the notion of a journey and inspires discovery and adventure. The play sequence is composed as a delicate insertion into the existing Melaleuca forest. A water theme, capitalising on the adjacent water quality central pond

and the nearby Haslams Creek, was extended through the playground by a series of shallow pools and channels.

The design of playground is extremely sensitive to the ecology of the site both functionally and aesthetically. It creates a popular adventure play area in a naturalistic setting, inviting families to explore the bush setting in a part of Sydney with limited access to nature. The play area complements existing overused play facilities in other parklands precincts and provides more challenging play experiences for older children and imaginative play opportunities for younger children.

Several paths draw children into, through and beyond the playground. The play elements include a series of specially designed finely crafted sculptural timber "boxes" that respond to the character of the site from the enclosed Casuarina and Melaleuca forest to the open expansiveness of the wider park. They mimic the journey from the forest to the park and include an enclosed bird hide, a "tunnel" through the forest, an unfolding, oversized sun-lounge, and a deck platform open to the landscape. A playful sense of scale also contributes to the richness of the play experience, as some of the box elements can be seen as oversized furniture elements. The scale of the playground, the multifunctional quality of the different elements and their integration with the natural setting is unusual in playgrounds and sets Wentworth Common apart.

The use of natural and recycled materials and finishes, detailed to reflect the patterns and textures of the surrounding heaths and forests helps to create a legible design language. With the growing Melaleuca "tunnel", and the forest paths, the boxes, extend the theme of a play journey through a series of outdoor rooms.

A playful sense of scale also contributes to the richness of the play experience, large climbing structures are perched in seas of sand, and a kinetic sculpture floats high above the playground.

This art work celebrates the site's natural setting; known as "The Birds" it is designed by Louise Pearson of HASSELL.

It is inspired by the majestic motion of wetland birds in flight and features a series of abstracted brass birds which perch on a wire over the playground. Their origami-like form is designed to move in the wind - subtly oscillating in a gentle breeze and becoming more expressive in a gusting wind. The art work was designed in unison with the design of the park to provide a truly site specific and integrated artwork.

Through rhythms of changing scale, light, shadow and enclosure, the playground invites children to engage with the site and see the playground afresh each time they visit. Natural materials have been used throughout the park, adding to the harmonious setting. In line with sustainability principles, recycled timber and reused sandstone are also featured across the site in order to create and extend the contrasting textures present in the melaleuca forest and native grassy landscape.

Wentworth Common is an important recreation facility and is just one key element of the legacy of the 2000 Sydney Olympic Games. Penelope Figgis AO, Chair of the Parklands Advisory Committee said; "The Park is a wonderful metaphor for the changing values of our society... the whole site is representative of the cumulative abuses of the land the ignorance of previous generations, a massively abused site. It is a metaphor for what we have done to the world as a whole. The Sydney Olympic Games and the parklands are a symbol of the changing attitudes towards the land... the need to protect it, restore it, heal and make it into a place of value and support for the community. It is proof that nothing is so damaged that it can't be repaired."

功能与形式的完美结合 —— 溪畔线性公园

Harmony between Aesthetics and Function — Creekside Linear Park

撰文：Cathy Wei　图片提供：HLA

该项目位于加利福尼亚州萨克拉门托市的北纳托莫斯区，是一个包括一条二级排水渠的开放型绿色走廊，设计成功地将排水功能和休闲功能完美地结合在一起。

背景

北纳托莫斯区是萨克拉门托市迅速发展的地区。在这一快速增长的环境里，占地面积 930 777 ㎡的溪畔社区内混合用地包括：近 1000 户独栋住宅和 450 户多层住宅、营业面积为 31 587 ㎡的社区商业中心、面积为 28 328 ㎡的社区公园和一所面积为 80 937 ㎡的幼儿园至八年级的校园。一个面积为 809 371 ㎡的大型公园未来将在溪畔社区的东面建成。该大型公园的东面和南面毗邻新建的高中、社区学院和公共图书馆。

占地面积为 17 401 ㎡的溪畔线性公园是溪畔社区和大型公园之间的线性边缘。

设计的主要内容包括：总体景观规划设计指导；公园、开放空间和街道等景观设计；商业中心景观设计；社区营销标志以及大部分社区内二级地段的景观设计。景观设计师在最初的开发阶段就参与到规划设计过程中，这对溪畔社区发展带来了特别的优势，整个社区以景观设计为重点，保护了社区内的自然生态环境，公园和游憩区便利可及，景致宜人、环境优美。

功能

该项目有两个功能：排水渠和步道与自行车道。排水管理是景观带很重要的功能，它也是复杂的排水系统的一部分。整个排水区包括总体规划的社区和未来的区域公园。除排水功能外，场地内复杂多样的公共设施也需要加以考虑。公共事业部门对有关排水管理和长期维护的基本要求是：沿公园需有花岗岩沙地和无需修剪绿地、种植本地草木的水渠斜坡和自行车通道。

景观设计师在不影响排水和公共设施功能的前提下，增强景观设计的美观和实用性，提高社区形象，使社区在总体上增值——溪畔线性公园形成绿色走廊，连接了社区内的住宅和商业中心，为学生们提供了步行上学的安全空间，扩展了城市总体规划的自行车道系统，可作为将来的区域公园的大门，同时为居民创造了美观舒适的休憩环境。

相对于以前，现在的环境发展密度更高，绿地往往首先成为被减少的用地，但充足的开放空间是一个社区生活质量的保障。像该项目这种把简单的功能区加以美化并与绿地开放空间相结合，是保证绿地面积的最佳方法。植物和硬质景观的布局都不影响地面排水的要求，重型的工程车辆可以随时使用景观带内的公共设施。该项目充分展示出景观设计为社区总体质量的提升所带来的价值。

设计

景观设计师的设计目的是创造一个充满魅力的、对生态友好的线性公园，对市容及市民活动空间产生积极的影响；减少用水、能源投入和绿色垃圾；增加一条重要的自行车和行人通道；同时通过在排水管道上方建造两座大桥与待建的区域公园相连接。

整个景观设计采用简单的软、硬材料，依靠集结、颜色和纹理提供视觉冲击。开阔的

溪畔居住区地图

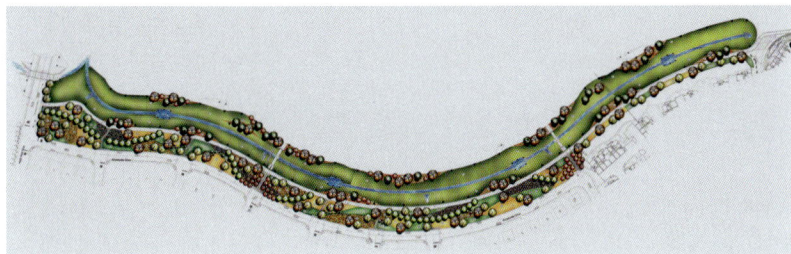

溪畔景观带平面图

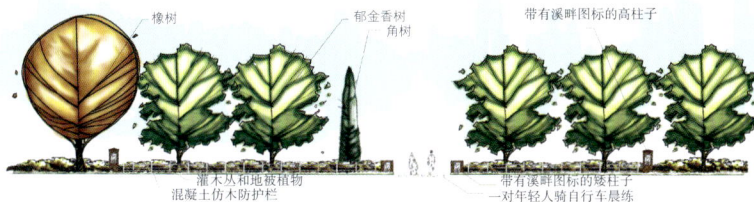

橡树
郁金香树
角树
带有溪畔图标的高柱子
灌木丛和地被植物
混凝土仿木防护栏
带有溪畔图标的矮柱子
一对年轻人骑自行车晨练

溪畔景观带入口立面图

空间由有节奏感的弧线分隔成种植区、花岗岩沙地区和卵石区（这些硬质材料都从当地获得）。加利福尼亚州各类本地草木随风飘动，为这处有韵律的设计增添了动感。树木的选择则考虑到建筑形态，增加了开阔空间中的垂直元素。由于地下公用管道设施众多，每棵树木的位置选择都是一个挑战。

该项目入口标志的设计延续了溪畔纳托莫斯总体规划中的纪念碑所选定的草原式建筑风格。在社区的其他开发项目中，同样大量采用了这种建筑风格，使整个社区浑然一体。

低用水、低维护、低废物

该项目采用独特的本土植物和天然材料来满足公共设施部门的主要设计原则，使景观具有蓄水功能且维护费用低廉。

植物的选择对实现上述目标是至关重要的，耐干旱的本土植物加上少量耐旱的观赏树种是极佳的搭配。观赏树种临近居住区街道的一面，多采用与周围街景相近的、相对葱郁的植物材料，景观带内大部分为极耐旱的本土植物。

中央控制灌溉系统在不同种植区设有不同的浇灌时间和强度，既能满足不同植物的用水需求又节省用水。在斜坡上设立的临时灌溉系统用以灌溉本地草本植被，将在成熟后移除，以进一步减少植物用水。草木灌溉采用地下深水浇灌，以避免地面喷灌系统挥发浪费的弊端。所有植物几乎不需要修剪，从而减少了绿色废物的产生和景观的维护费用。

1　向西鸟瞰溪畔景观带

2　人行道旁的坐椅

3　溪畔景观带中的植被、卵石和花岗岩沙地相互交错

4　小型步行桥跨过排水渠连接溪畔居住区和待建的区域公园

Creekside Linear Park is an open space corridor located along the secondary drainage channel in North Natomas area in the City of Sacramento, California. This linear park successfully combined storm water management functions and recreational greenbelt in an aesthetically pleasing way.

Background

North Natomas is a rapidly growing area in the City of Sacramento. Within this fast growth, the 230-acre Creekside community composes of mixed-use developments of nearly 1,000 single-family homes and 450 apartments; 340,000 square feet of neighborhood retail; a 7-acre neighborhood park; and a 20-acre K-8 school campus. A 200-acre regional park is proposed to the east of the community, adjacent to a newly constructed high school, community college and public library.

This 4.3-acre Creekside linear park forms the edge between the Creekside community and the grand regional park.

The master developer, Lewis Operating Corp., started developing the Creekside community in 2001. The HLA Group, Landscape Architects and Planners, led the planning and design efforts from the start, services included landscape design guidelines for the master planned community; landscape design for the parks, open space, streetscapes, and commercial town center; the community's marketing logo; and design services for majority of the merchant builders developing the different parcels. Being involved from the initial planning and design stage gave the landscape architect special advantages in creating a landscape emphasis on the whole community. A landscape based planning and design results in ecologically friendly environments, easy access

to parks and recreational areas and aesthetically pleasing surroundings.

Function

The Creekside linear park has two functions, drainage channel and pedestrian / bike path. Storm water management is the primary function of the site, which is part of a complex system of channels that drain the master planned community and the future regional park. Access to varied utility structures along the park also needs to be considered. The Utility Department requirements related to functional storm water management and long term maintenance were basic: decomposed granite and no-mow grasses along the utility corridor, channel slopes planted with native grasses, and a linear bike trail.

The landscape architect's approach was to augment the utility corridor with landscape enhancements that create a community amenity without interfering with the functional requirements of storm water management. The result is a green corridor that connects residential developments, links to the commercial Town Center, provides off-street access to the high school, extends the City's master planned bikeway system, serves as a gateway to the future regional park, and creates attractive views for residential development.

The utilitarian functions of storm water management and access were designed in a manner that created a memorable place that gives people access to natural open space in a higher density suburban environment where every bit of open space counts. The placement of plant and hardscape materials does not interfere with the overland release requirements for neighborhood storm water drainage. Heavy utility vehicles can readily access all the utilities running under the length of the park. Elevating a functional utility corridor into a linear park amenity for a new mixed-use community shows the value landscape architecture adds to development that embraces the principals of smart growth.

Design

The landscape architect's design intent was to create an attractive, ecologically sound linear park that serves as an active amenity; reduces water usage, energy inputs and green waste; functions as a major bike and pedestrian corridor; and links to the future regional park via two bridges over the drainage channel.

With a simple palette of soft and hard materials, the design relies on massing, color and texture to provide visual stimulation. A sense of rhythm was introduced to the linear site through the geometry of interlocking circular segments that delineate areas for planting, decomposed granite and cobble (both locally obtained). Grasses and sedges indigenous to the California landscape move with the wind to add a kinetic element to the site's rhythmic design. The tree palette was selected for its architectonic quality, adding a vertical element to the linear plane. Canopy trees include evergreen oaks—which are icons of California's Central Valley. Positioning trees to support the park's overall design was a particular challenge, given the extensive amount of underground utilities.

Entry elements connecting the park to the adjacent neighborhoods reflect the Prairie Style design used for monuments within the Creekside Natomas master planned community. Using an iconographic landscape and built elements from the community seamlessly integrates the utility corridor into the larger suburban environment. These design and plant elements are being used by other master builders working within the region to extend the vernacular and further unify the new development.

Low Water Usage, Low Maintenance and Low Waste

Creekside Linear Park uses a unique application of native plants and natural materials to meet the Utility Department's primary design goal of providing a low maintenance landscape with water conserving principles.

Plant selection was critical to realizing these goals. Native plant materials were mixed with non-native ornamental species with low water requirements. The ornamental species used along the park edge that fronts the houses relates to the context of residential planting, creating a green front door that is more readily acceptable to the public. Throughout most of the park, native sedges and fescues were alternated within ribbons of hardscape materials carefully set in gently undulating patterns.

The centrally controlled irrigation system is hydro-zoned to satisfy the differing water needs of the selected plants. Supplemented irrigation for the native grasses planted on the channel slopes will be removed once the native plants are established. Trees are deep water irrigated to eliminate water waste compared to above ground spray systems. All of the plants require minimal pruning, which reduces green waste and minimizes the cost of on-going maintenance.

1   人行道可以通到当地的学校
2   人们惬意地在溪畔景观带上行走和骑车
3   溪畔景观带中大量使用了本地采集的卵石
4   溪畔景观带北端入口

莫伊拉河滑板运动公园

The Moira River Skatepark

撰文：Jim Barnum(Spectrum Skatepark Creations Ltd.)　　翻译：王玲

1　碗形滑板场地俯视图（图片提供：Brent Jordan）

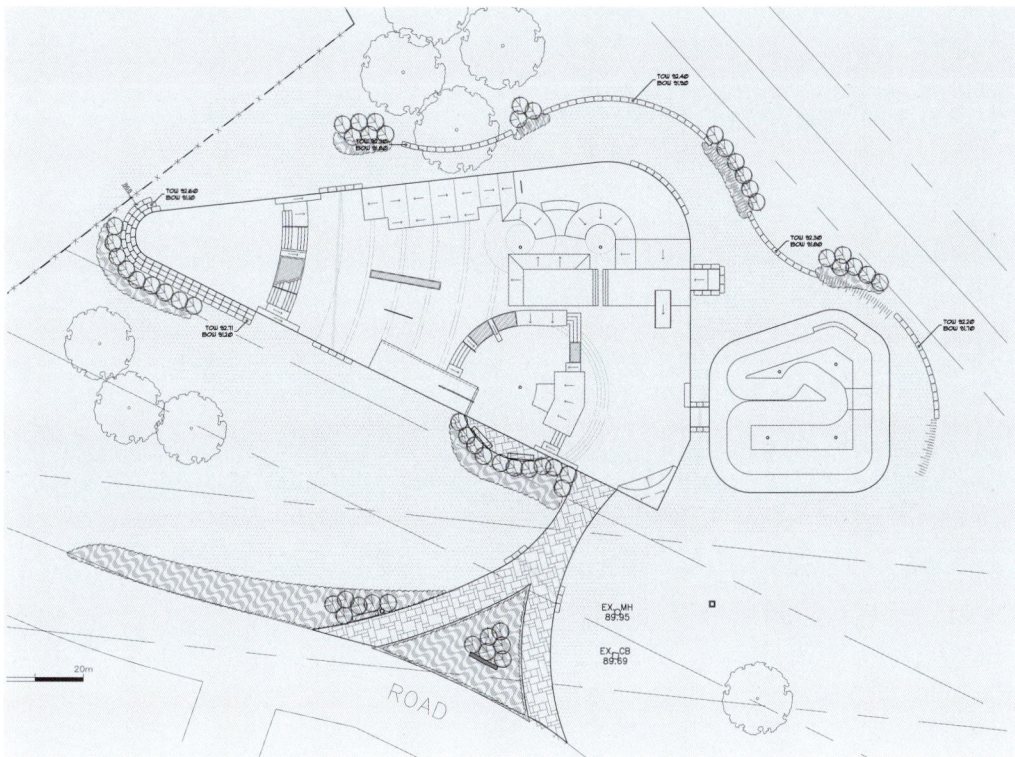

总平面图（图片提供：Bill Gurney,LAND Inc.）

莫伊拉河滑板运动公园和广场是一处专为滑板爱好者设计的、富于动感的硬质景观。滑板爱好者是一个容易被忽视的、处在社会边缘的群体，因此规划和设计好滑板运动公园将对社会产生重要而又积极的影响。设计团队积极邀请当地青年参与到滑板运动公园的设计中，其中许多功能性设计方案都直接遵循滑板爱好者的意愿而设计。为了确定贝尔维尔市滑板爱好者心目中的公园形象，设计团队举办了一场由滑板爱好者参加的设计研讨会，请他们画出或者描绘出自己喜欢的公园元素以及滑板运动所需要的跳跃和坡道设施等。设计团队邀请当地滑板爱好者参与到规划中，并认真聆听、分析他们的意见，增强了滑板爱好者的自我价值感和归属感。贝尔维尔市也支持设计团队的这个做法，并积极配合使之推陈出新。设计团队与贝尔维尔市共同把握住这次机遇，关注那些常常被忽视的青年群体，鼓励他们融入社会并与市民共同参与其中，而不是对他们置若罔闻。因此，公园不仅成为这部分青年群体释放活力的安全场所，而且也为他们经常表现出的消极行为提供了一种替代方式。

为了推动滑板运动公园的设计在社会层面上进一步发展，占地1470㎡的莫伊拉河滑板运动公园不仅体现了滑板爱好者的愿望，同时也展现出其所应承载的社会功能。公园的设计灵感来自附近的莫伊拉河，体现出了对自然环境的无比尊重。莫伊拉河是大多数加拿大人心目中的钟爱之地，几乎所有的附近居民都对自然心存感激，因此该项目的设计目标是使贝尔维尔市的滑板爱好者更多地认识和欣赏这条美丽的河流，对其产生由衷的自豪感，并在这些看似"局外人"的滑板爱好者之间创建共同点，使他们拥有与贝尔维尔市其他市民可以共同分享的财富——对自然美景的自豪感。

莫伊拉河十分独特，当它流经贝尔维尔市中部地区时，水流会变得错落有致，跌落的叠层小瀑布仿佛加拿大前寒武纪地盾岩层上的立体雕塑一般。滑板爱好者喜欢台阶和落差，因此错落的河流成为了滑板运动公园功能设计首选的借鉴元素。公园内宽敞的开阔空间象征着绵延宁静的河滨地区；台阶和高差的突然变换将公园分成不同的区域，它们与河流中的叠层小瀑布相得益彰。

"流畅性"是所有滑板运动公园设计成功的关键，因为流畅性意味着道路的不间断性以及获得和保持速度所需要的最小推力。没有什么比水更富灵动性，因此从水中寻找设计灵感也是设计师必然的选择。除了错落叠层的大手笔设计，公园的功能性设计还包括一些特殊的滑板技巧训练设施，它们仿佛"翻滚的波浪"和"流畅的弧度"，淋漓尽致地展现出水的特质。形式服务于功能，因此视觉上的流畅性也完美地展现出滑板运动带给人们的畅快淋漓的感受。

为了使公园在贝尔维尔市独具特色，设计师运用了许多滑板公园未曾使用过的创新元素，如"三台阶平滑高差变换"和"平滑向上前轮离地的组合动作"的设计。

从功能性设计到美观设计，滑板运动公园包括一系列"锯切"的带状设计（在工程二期，它们将被酸蚀刻以增强视觉冲击力），仿佛荡漾的水波，悠悠地沉入公园。公园最低处的"锯切"状设计是一个平滑的"水池"，设计灵感来自于石头打在水面上泛起的层层涟漪。

公园中应用的钢材主要用于保护混凝土的边缘，并采用耐磨防锈漆将它们涂成蓝色。

　　滑板运动公园及其周围的多功能设施吸引着各个年龄层的人来此休闲放松。场地富有社会敏感性，这也正是将贝尔维尔市自然美景融入设计的另一个原因。正如滑板爱好者在参与设计过程中更加融入社会一样，整个社会对这些滑板设施也逐渐接受，从而对滑板爱好者更加认可，因为公园不仅仅是为了满足滑板爱好者对其功能的需求，它触及了更加深刻的社会层面。公园的设计尊重了贝尔维尔市的所有市民，不仅美观实用，而且充分展现出贝尔维尔市民所热爱的当地的自然美景。

The Moira River Skatepark and Plaza is a dynamic hardscape designed for skateboarders. Skateboarders can often be somewhat disenfranchised youth on the fringes of society, so there is significant potential for a positive social impact through good planning and design of skateboard parks. Local youth were heavily included in the design process, with most of the functional design program being directed by their wishes. To determine what the skaters of Belleville wanted to ride, we held a dynamic design workshop to which the skaters were invited and asked to sketch and describe what kind of elements, jumps, ramps etc. they would like in their park. By including local youth in the planning process and actively engaging & listening to them, we increase their sense of self-worth and belonging. The City of Belleville agreed with our inclusive design process and leveraged it to make their skatepark project turn out much more than just a recreation facility. Together we seized the opportunity for their community to reach out to an often disenfranchised segment of youth and involving them in the community and civic processes, instead of pushing them away. By providing a safe, positive place and an outlet for this underserved segment of youth, we have provided an alternative to the negative behaviours often displayed by disenfranchised adolescents.

Aiming to take the social aspect of skatepark design a step further, the 15,780 square foot (1,470 square metre) Moira River Skatepark was designed to reflect not just the wishes of the skaters, but something of the character

of the community in which it was to sit. The design was informed by the nearby Moira River, honouring the natural environment, which is something dear to the hearts of most Canadians. Appreciation of nature is something that's shared by nearly all people, so our goal was to make the skateboarders of Belleville more aware, appreciative and proud of their community's beautiful river. Thereby we have created common ground between the "outsider" skaters; we have aimed to give them something that they can share with the rest of the citizens of Belleville: pride in their beautiful natural environment.

The Moira River is a unique water course. As it flows down the middle of Belleville, much of it displays a stepped configuration, cascading down small waterfalls that it has carved out of the pre-Cambrian rock of the Canadian Shield. Skateboarders love steps and drops, so this aspect of the river was an easy choice to incorporate into the functional design of the skatepark. It is expansive & fluid with wide open spaces reflecting the long, tranquil areas of the river, punctuated into different zones by sudden elevation changes of stairs and drops for the skaters to launch down; like the river's waterfalls.

The aspect of "flow" is critical to the success of any skatepark. In terms of skatepark design, flow refers to ensuring that paths of travel are uninterrupted, and that a minimum of pushing is required to gain and maintain speed. Nothing is more fluid than water, so nothing was a more fitting inspiration for this park. In addition to the stepped, cascading big picture design, the functional design also includes specific skateboard features that are undulating and flowing such as the "roller waves" and the "flowing hubbas", to reflect the nature of water. Function follows form, and thus the elements that are fluid to the eye are also perfectly fluid for skateboarding.

In an effort to make the park even more unique and truly Belleville's own, it includes some brand new, never before seen skatepark elements such as the "three stair flat gap" and the "flat to up many combination units".

Moving from the functional design to the aesthetic, the skatepark includes saw-cut banding (which, in a second phase of construction, will be acid-etched to increase the visual impact) which ripple out like waves from the first, and most major, drop in the park. Additional saw-cut patterns in the lowest area of the skatepark, "the pond", were inspired by the ripples that emanate from a stone tossed into a pond. The steel "coping" in the park which is required to protect the concrete edges from the skateboarders and their tricks was painted with durable anti-rust paint in blue.

The skatepark had to fit into a parkland site with multiple adjacent uses that attract users of all ages. Thus the site is socially sensitive, and this was another reason to include design aspects that respect the beauty of Belleville. Just as the skateboarders feel more drawn into their community by being included in the design process, the community-at-large has also been much more accepting of the facility, and thus the facility users, due to the fact that the skatepark reaches beyond just the functional needs of the skaters. It respects all citizens of Belleville by being aesthetically pleasing and reflecting the beauty of the area that all citizens of Belleville share and love.

1　滑板爱好者在表演（图片提供：Brent Jordan）
2　壮观的入口处（图片提供：Jay Bridges, BSE）
3　滑板爱好者集会（图片提供：Jay Bridges, BSE）

儿童乐园

Guys Garden

撰文：Aaron Weingrod & Michael Abrahamson   图片提供：Tzur Kotzer & Michael Abrahamson   翻译：谷晓瑞

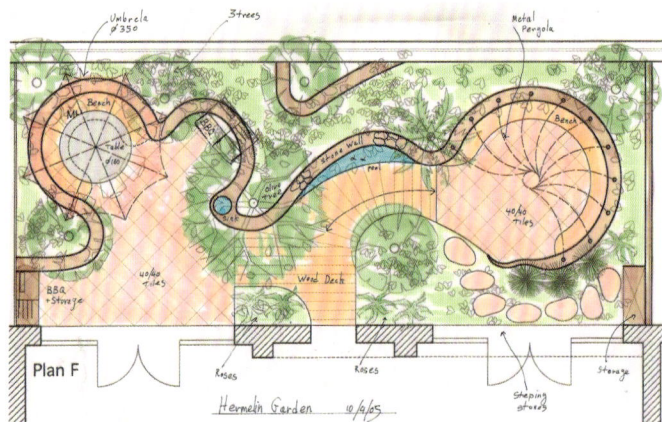

花园总平面图

设计公司在设计前通常会先与客户交谈，了解他们期望得到的效果，在项目中孩子们给设计师带来了渴望已久的灵感。设计师让孩子们将理想中花园的样子用彩笔画出来给他们看，画面中有流水、隧道、山洞和情景，竟全然没有一条笔直的线条或90度的转角！Weingrod-Abrahamson 建筑公司决定以这些为基础打造一个充满孩子气息的花园，同时充分利用原有条件并满足花园的使用功能。这座位于市区的花园长 16m、宽 5m，后面是一道高 2m 的围墙，两侧还有其他花园，这些都给设计带来了很大的挑战。

设计师决定要使这个美丽但植被稀少的戴维斯村庄回归自然。设计师用露台上粗糙的石头筑起了一道蛇形的围墙，石头透着微红、灰白、铜绿的本色，不禁使人联想起地中海地区常见的圣经露台。这道围墙是花园的主干，蜿蜒的空间延伸形成了休息区、鱼池、烧烤区和早餐区，另外还设置了一个传音筒供孩子们窃窃私语。茉莉花和紫藤花不久后便会爬满金属支架的凉亭，那时这里将香气宜人、无比阴凉；而凉亭后侧的那些大树刚好遮住了简陋的工作场所。与隔壁相邻的围墙是用当地一种薄薄的石子筑起的石墙，在其内侧镶了一层桉木墙，从而保持了材质和色彩的自然效果。

改造后的花园开始给人们的第一感觉是一个独立的小空间，但走进其中便会慢慢发现这是三个各具特色又和谐统一的广阔空间。

WAA always begin their dialogue with clients by asking them their wish list for the new design. The children gave them the inspiration they were looking for. The architects were presented with many colored sketches of a child's fantasy garden, completed with water, tunnels, caves and drama. Not one straight line or 90 degree corner! With this as their starting point, Weingrod-Abrahamson Architects tried to design a garden with the children's spirit, while encompassing all the conditions of the site and "functional" needs of a garden. This urban site, being a level 5X16 meters long, with a 2 - meter wall at the back, and neighbors gardens on both sides—provided a real challenge!

WAA decided to bring back nature to this "beautiful" but sterile David's village, where the average stylized apartment costs over 1 million dollars. They designed a snake wall of rough terrace stones with the original reddish/grey patina…reminiscent of the biblical terraces one sees in the local Mediteranean landscape. This wall was the backbone of the garden moving up and down and around to create sitting areas, a fish pond, barbeque area and breakfast niche. Even a "speaking tube" for the children to whisper into was added. A metal gazebo which will soon be covered with jasmine and wisteria will produce both scent and shade. The larger trees at the back hide an ugly work site. The wall to the neighbors was lined by a thin tile of local stone with a split eucalyptus tree plank wall on the inside, continuing the natural effect of material and color.

Upon completion, one feels that the single small space at the start, grew into 3 large separate spaces each with its own identity but harmoniously linked together.

Northala Fields 公园

Northala Fields

撰文：Jasmine Ong　　图片提供：FoRM Associates　　翻译：董桂宏

1　具有象征意义的土丘

2　航拍图

3　2005 年施工中的场地

总平面图

　　该项目位于诺索尔特·格林福德郊野公园（郊野公园位于伦敦西部地区，占地面积100万平方米）的腹地，占地面积185 000 ㎡。该项目在保留了郊野公园原有设施及特色的同时，使公园设施更加完善、特点更加鲜明。

　　经过多方数年来的讨论，设计方案终得以出炉。该项目最大的特点是改变了这一地区的地貌特征：设计师特别从里斯本进口了大量的建筑用鹅卵石（这种鹅卵石也被广泛应用于 Terminal 5 和温布利竞技场等项目）。全新的地貌解决了场地发展的一系列问题：

　　·降低了邻近高速公路的噪音污染、视觉污染和空气污染等对场地的影响；

　　·增设了全新的娱乐设施，但目前新的娱乐设施尚未推广到整个郊野公园；

　　·具有生态保护作用，至今已减少了13万辆运货车驶过伦敦郊区；

　　·作为重要的"土地艺术"作品，该项目已成为了伦敦西部入口的地标。

1

新地貌的主要特色是四座圆锥形土丘，它们坐落于场地的北部，海拔分别为 20m、25m、30m 和 35m，土丘坡度大约为 30 度左右。

一系列泾渭分明的道路保证了公园的娱乐功能。主要道路与二级道路形成网格状，连接着周围广阔的空间，这也是郊野公园的另一特色。全新的游乐场位于公园的主轴线上，游乐场两侧绿草青青，并且设有适合沉思冥想的区域，满足了不同年龄使用者的需要。场地的中心湖兼具环保功能与娱乐功能，作为伦敦惟一的城市公用渔场，中心湖是人们休闲垂钓和划船戏水的理想去处。

该项目的第三大特色是环保功能：

· 林地——设计师增加了原有林地的密度，并且尽量利用原有的生态资源，如将场地内枯死的榆树篱应用于新生林地之中。

· 草坪——绿草是场地内的主要植被，青绿的小草铺满四座圆锥形土丘和场地南部，一些国外的植物也为场地增色不少。如今，这里已成为动植物群聚集的生态展厅。

· 水与湿地——新建的水道有利于水生动植物群落的繁衍生息。

水是舞动的精灵，也是娱乐功能与生态功能的核心元素。全新的公园水系统有效地提高了地表水与地下水的循环利用，通过在地表凿洞源源不断地获取地下水，而地下蓄水层又保证了地下水的供给。这一水系统保证了在少雨季节里城市公共渔场的水位高度。

1　主游乐场
2　百花坡
3、4　填石铁笼墙

1 植物景观

2、4 远景

3 土丘漫步

Northala Fields 18.5 hectares of open space lie at the heart of the Northolt and Greenford Countryside Park, a network of open spaces covering 100 hectares in West London UK. The development of Northala Fields connects existing facilities and features within the Countryside Park and provides new facilities and features that are not currently available to the local community.

The design proposal has been developed over a number of years through a comprehensive public consultation process. Arguably the most significant feature of the Nort hala Fields design is the construction of a new landform on the site, utilising substantial volumes of imported construction rubble onto the site from a pool of London wide development projects such as Terminal 5 and Wembley stadium. The deposition has successfully delivered £ 6.2 million income as the main funding of the park as well as significantly contributed to shrinking the ecological footprint of London by avoiding over 130,000 lorry journeys to outlying tips.

The new landform provides a solution to a number of site and development issues:

• Mitigation of impact from adjoining motorway, particularly noise, visual and air pollution

• Creation of new recreation opportunities not currently available elsewhere in the generally flat Countryside Park

• Creation of new ecological opportunities through new topography and soils

• Creation of a major piece of 'land art' that will become a landmark gateway for West

London The dominant feature of the new landform are four conical mounds located along the northern edge of the site. The highest point of each of the four cones is (from west to east) 20m, 25m, 30m, 35m above existing ground level with slope gradients of 1 in 3.

A series of clearly defined key routes support recreational uses and activities of the new neighbourhood park. A network of primary and secondary paths connects with adjacent open spaces that make up the rest of the Countryside Park. New playground, meeting the needs of a variety of ages and users, is set along the central spine of the park along with a series of open meadows and semi-formal planting and seating areas for more contemplative activities. The central network of lakes is seen as much as an environmental resource as a recreational one by providing opportunities for fishing as the only communal urban fishery in London and for model boating. Enhancing the ecological values of the site has also been a focus of the new design.

A range of new habitats was created:

• Woodland—both around the perimeter and within the site adds to the diversity of the existing woodland habitat. Where possible, the existing ecological resource remnant in the dead elm hedgerows within the site were salvaged and integrated into the new woodland areas.

• Meadow and grassland—meadow types are the dominant vegetation in the development, covering the proposed new cones and most of the area to the south. A range of meadow types was established through the careful selection and placement of imported material and differing management regimes, providing an ecological showcase for different meadow and grassland flora and fauna.

• Water and wetland—the new watercourses provides opportunities for water and wetland flora and fauna that are not currently present on the site.

Water is at the heart of the reacreation and ecological gains of the new design. The new water system captures ground and surface water drainage from the new landform and toped up from a borehole supplying groundwater from a deep aquifer. The borehole allows a constant water flow to be maintained through the water system, particularly during periods of low rainfall to ensure sufficient water levels in the urban fishery.

五角大楼纪念公园
The Pentagon Memorial

撰文：Lee+Papa and Associates　　　图片提供：Craig Atkins　　　翻译：申为军

　　2008 年 9 月 11 日，斥资 2000 万美元的五角大楼纪念公园开始对公众开放。为了纪念 2001 年 9 月 11 日五角大楼遭遇的恐怖袭击事件以及五角大楼和美国航空公司 Flight 77 航班中的 184 位遇难者，将纪念公园建在了五角大楼西南面的一块占地 8093 ㎡ 的土地上。纪念公园没有给参观者灌输固定的思维或感受，而是鼓励参观者自己进行思考，用自己的方式去铭记和回忆那一天的事件。在体验方式上，该项目打破了自然地理环境的束缚，使那些驾车或乘坐飞机经过此地的人们也能见证其独特的布局和灯光设计。同时，纪念公园也为遇难者亲友及其他团体提供了一个宁静的悼念空间。

　　五角大楼纪念公园由遍布公园的 184 个纪念碑依次排列而成。这 184 个纪念碑实际上是一系列的悬臂式长凳，悬臂下是一条流淌的小河。纪念碑有着优雅的自身支撑形态——既是可以发光的的荧光池，也是悬臂式长凳，上面镌刻着每一位遇难者的名字。每一块纪念碑都对应一位遇难者，并按年龄大小在场地上依次排开——从 3 岁的 Dana Falkenberg 到 71 岁的 John D. Yamnicky。以出生年份来确定年龄线，不锈钢数字被水平镌刻在场地边界长凳的抛光表面上，两侧的不锈钢年龄线贯穿整个场地。59 个纪念碑朝向同一方向，另外的 125 个则朝向相反的方向，用来区别美国航空公司 Flight 77 航班上的遇难乘客和五角大楼内的罹难者。当人们瞻仰五角大楼内罹难者的纪念碑时，会发现镌刻的姓名与五角大楼在同一视角，而 Flight 77 航班上的遇难者的姓名则与天空在同一视角。

总平面图

整个场地的步行区被设计成稳定性强的透水砂石地，场地周围的边界长凳被用做种植槽，里面种植了观赏草。踏在松散的砂石地上可以让人们倾听和感受彼此的脚步声，同时这种砂石铺装也具有一定的牢固性和通达性。沿公园边界长凳种植的观赏草形成一道柔和的屏障，划分出纪念公园的界限；边界长凳及纪念碑可提供超过 640 米长的坐位空间。纪念公园的入口通道建在边界长凳的外围，这样既保证了人们可以安静地入园，同时也充当了纪念公园与五角大楼的南停车场和华盛顿大街、安保通道之间的缓冲通道，可将那些较为喧闹的活动，例如旅游团的活动拒之门外。

场地的西部边界筑有年龄墙，其高度根据年龄而定，年龄每增加一岁就升高一英寸（1英寸等于2.54cm），并与整个场地的年龄线相对应。当人们向公园的深处走去，年龄墙也随之升高——从高出公园边界长凳3英寸（Dana的年龄纪念线）上升到18英寸高长凳（John的年龄纪念线）之上的71英寸高。从整体规划来看，这面墙是纪念公园与位于公园西北侧的安保通道之间不可或缺的屏障。在公园中，年龄墙也为那些生长在底部的观赏草提供了适宜生长的阴凉环境；而纪念碑旁边也都种植了树木，因此场地内部也不乏阴凉之处。

凯斯·凯斯曼（Keith Kaseman）和茉莉·贝克曼（Julie Beckman）紧密合作，通过细致的研究和对细节的关注，形成了该项目完美的设计理念。公园的设计和建设不但符合遇难者家属的愿望，并且在一定程度上，公园的建立也与五角大楼的历史和自然因素相适应。在公园的规划过程中，五角大楼纪念公园基金会代表了罹难家属的意愿，并为公园的建设募集了必要的资金。

Lee+Papa设计事务所与鲍佛·贝蒂建筑公司合作，成功地完成了纪念公园的设计和施工。作为该项目的主要设计单位，Lee+Papa充分发挥了设计的领导才能，统领着来自全国各地的各行业设计师，如概念设计师、工程师、水景及灯光顾问等，向世人展示了文化资源和景观设计的新成就。

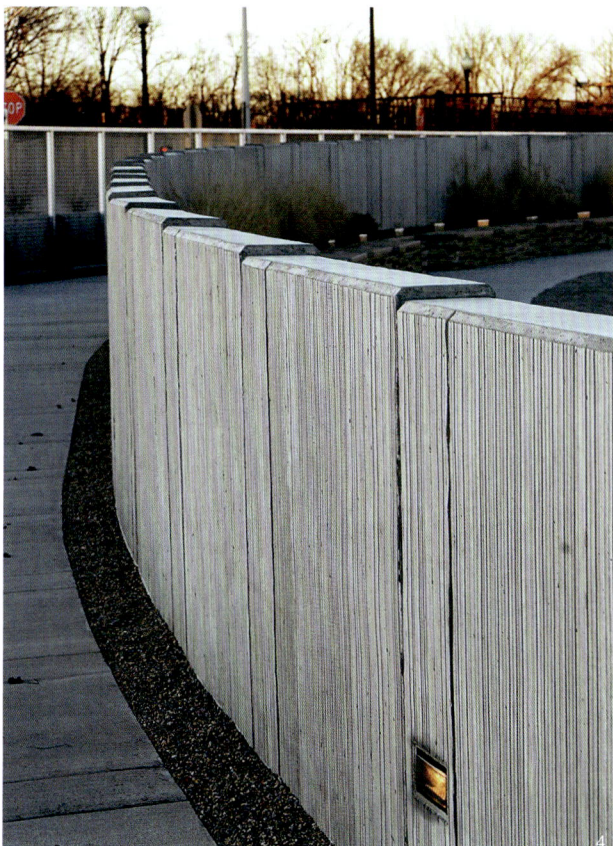

1　不锈钢镌刻
2　前来悼念的游客
3　夜景
4　年龄墙
5　出生年份

The Pentagon Memorial, a 20 million dollar project, was opened to the public on September 11, 2008. The memorial, built on a 2-acre site at the southwest facade of the Pentagon, commemorates the September 11, 2001 terrorist attack on the Pentagon and the 184 lives lost in the Pentagon and on American Airlines Flight 77. The park is meant to be experienced in a non-prescriptive way, allowing visitors to interpret the space on their own and remember and reflect on the events of that day in their own way. The park experience extends beyond the physical constraints engaging those driving by or flying over witnessing its unique layout and lighting scheme. The memorial simultaneously affords intimate and collective contemplation through silence within a tactile field of sensuous experience.

The memorial park is centralized around the 184 inscribed memorial units disbursed across the park. The 184 memorial units themselves are a series of cantilevered benches with a glowing pool of constantly circulating water beneath. Elegant in its self-supporting form, the Memorial Unit is at once a glowing light pool, a cantilevered bench and a place for the permanent inscription of each victim's name. Each unit is dedicated to an individual victim. The field is organized as a timeline of the victims' ages, spanning from

Dana Falkenberg, 3 years old, to John D. Yamnicky, age 71. Birth years, used to locate the age lines, are inlaid stainless steel numbers set flush with the finished horizontal surface of the perimeter benches. The birth years are flanked by the stainless steel age lines that permeate the whole site. Fifty-nine Memorial Units face one direction, 125 face the other, distinguishing between victims on board American Airlines Flight 77 from those inside the Pentagon. When visiting a memorial dedicated to a victim who was in the Pentagon, the visitor will see their engraved name and the Pentagon in the same view. Conversely, one would see the engraved name of a victim on Flight 77 with the sky.

The entire site walking surface of the park is a porous stabilized gravel field contained within two perimeter benches that serve as planters for ornamental grasses. Though loose enough to hear and feel footsteps upon it, stabilized gravel is a hard, accessible surface. The grasses along the perimeter benches act as a soft screen demarcating the boundary of the memorial park. The combined length of the perimeter benches plus the bench portion of each Memorial Unit provides more than 2,100 feet of seating surface. A Memorial Park Gateway was constructed on

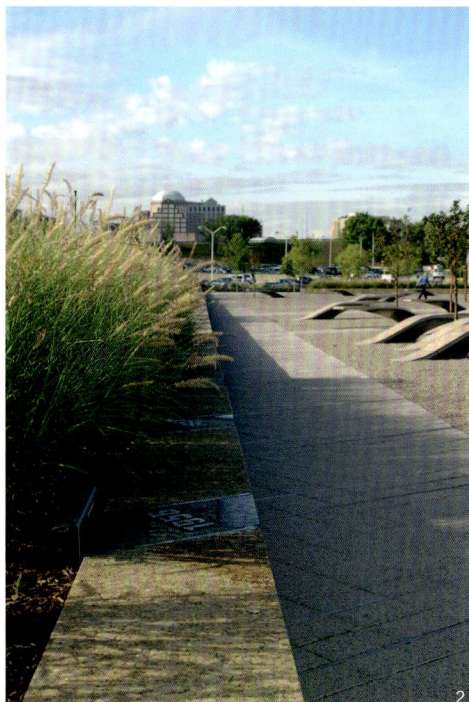

the outside of the perimeter benches to provide a peaceful entrance to the park. The Gateway serves as a buffer between the Memorial and both the Pentagon South Parking Lot to the south and Washington Blvd and the Secure Access Lane to the west, and removes some of the more intrusive activities out of the park such as tour group convening.

The western edge of the site is defined with the age wall — a wall that "grows" in height one inch per year relative to the age lines that organize the site at large. As one moves deeper into the site the wall gets higher — it grows from three inches above the perimeter bench (at Dana's memorial age-line) to seventy-one inches above the eighteen-inch tall bench (at John's). Strategically, this wall grows as a barrier and is needed between the memorial park and the secure access delivery lane that encroaches upon the site at its northwestern edge. From within the site it provides a shadowy backdrop for the lacy ornamental grasses that are planted along its base. Dappled shade is available around the site, as trees are clustered in conjunction with the disbursement of Memorial Units.

The seamless beauty of Keith Kaseman and Julie Beckman's winning concept for the Pentagon Memorial

was realized through intense collaboration, research, and attention to detail. The design and construction of the memorial was undertaken with great sensitivity to the families' wishes and integration of the memorial in a manner appropriate with the historical nature of the Pentagon. The Pentagon Memorial Fund represented the families' interests in the planning process and are raised the funds necessary to build the Memorial.

Lee+Papa formed a joint venture with Balfour Beatty Construction Company and successfully competed to be chosen as the Design/Build contractor for the memorial. As architect of record for the project, Lee+Papa provided design leadership, managing an interdisciplinary team from all across the country including the concept designers, engineers, pool and lighting consultants, in addition to providing cultural resource and landscape architectural expertise.

绿色新空间 —— Turruwul公园

New Green Space — Turruwul Park

撰文：HASSELL　　　图片提供：Brett Boardman　　　翻译：董桂宏

1　木栈道
2　避开树根区的混凝土道路
3　广受欢迎的烧烤设施

罗斯柴尔德大道

哈考特广场

海耶斯路

樱草大道

塔拉根自然保护区

AMENITY HUB

PLAYGROUND

总平面图

1　巨大无花果树荫蔽下的售货亭和更衣室
2　建筑均由上方的木板遮蔽，形成统一的整体
3　地面雨水可以通过草坪进入雨水回收系统
4　"之"字形造型的木质幕墙设计，高低起伏、线条流畅

注
1. 更衣室
2. 公厕
3. 售货亭
4. 烧烤区和阴凉区
5. 寄存处和饮水区
6. 网球场
7. 自行车道
8. 无花果树

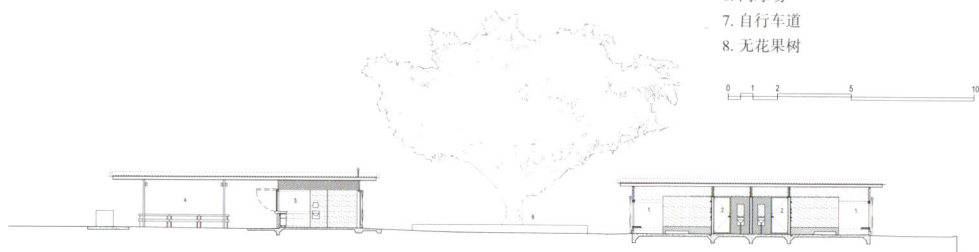

景观建筑剖面图

　　该项目位于悉尼市南部的郊区 Rosebery。悉尼市议会一向致力于公共公园的开发，并委托 HASSELL 公司全面改造 Turruwul 公园。公园内的无花果树枝繁叶茂、摇曳生姿，是公园最鲜明的特色之一。虽然公园缺乏总体布局——部分建筑群年久失修、娱乐设施陈旧、公共步道有限、缺少公共空间，但是其公共空间存在着巨大的开发潜力，这也是该公园的另一大鲜明特色。因此，该项目的主要目标是突出葳蕤的无花果树，并充分利用硕大的树木来营造景观。在设计过程中，公园

中两棵最大的无花果树成了点睛之笔，设计师还引进了全新的游乐场和娱乐设施，并通过移除杂物来扩展公园的公共空间。

通过征求当地社区居民的意见，设计师为改造公园提出了建设性的意见，并制订了该项目的总体规划方案。考虑到休闲娱乐的需要，设计采用了全新的基础设施和娱乐设施，包括建筑、道路铺装、体育设施、照明设施、景观装饰、观赏植物和游乐场地等。另外，设计师还充分考虑到家庭娱乐的需要，扩展了游乐场的面积，并采用了全新的枢纽，以使公园在空间上得到扩展，在品质上得到提升。在场地原有条件下，设计师注重公园的功能性和视觉性效果，巧妙地插入全新的设计元素以改善总体结构，使其成为同类改造项目的典范。经过改造，该项目成为该地区的核心休闲娱乐区域，是人们休闲娱乐的理想场所。全新的游乐场在一棵硕大的无花果树的荫蔽之下，由一系列活动区组成，而各个活动区之间由植物景观、卵石步道和木质步道相互连接。公园的体育设施也焕然一新，全新的运动场、网球场、板球场和篮球场满足了更多体育运动爱好者的需求。

设计师重新规划了公园内的道路（包括公园周边的林阴大道），使人们能够更好地利用公园设施，与大自然亲密接触。在施工过程中，设计师又遇到了一大难题，即如何在铺设新路的同时保护好地下细密的树根。设计师与悉尼市的园艺师们精诚合作，创造性地采用剪刀撑结构，并在表面铺设了木板路，这样既可以避免现有的树根受到破坏，又不会影响树根的生长。木板路不仅有利于空气流通和雨水渗透，还为树根生长提供了良好的环境，因此新铺设的道路完全不会影响树木的生长。

该项目还新增了售货亭和更衣室，为来此休闲运动的人们提供了便利。设计师独具匠心，在硕大的无花果树两侧建造了公共建筑群，并将这一区域改造成该项目的标志性核心景观。设计的画龙点睛之笔在于对公共建筑群木质幕墙的设计——木质幕墙的表面高低错落，"之"字形的造型仿佛正在被弹奏的钢琴键盘般高低起伏。虽然木料本身的纹理各不相同，但这种剪切设计体现出独特的设计工艺，为公园增添了独特的艺术魅力。设计师同时也非常注重环保，相邻的板条均取自同一块大木料，避免了木材的浪费。这组建筑群的设计简洁却不简单，具有很大相似性的建筑有机地排列在一起，为人们休闲娱乐提供了便利。

该项目的客户是位忠实的环保爱好者，主张依靠太阳能供应热水，并提供了许多太阳能光板；提倡自然通风、自然采光、使用环保材料、进行雨水回收与再利用等。公园内采用滴灌式的灌溉方式，有效地节约了水资源，网球场和道路上的雨水都能够沿着低洼地流入雨水回收系统。

Turruwul Park is located in Rosebery, a south Sydney suburb. The City of Sydney Council is committed to deliver innovative improvements to their public parks and HASSELL was commissioned to upgrade Turruwul Park. The defining character of the park is its beautiful established trees and open spaces which had become lost behind tired recreational facilities, a dilapidated cluster of buildings and a lack of overall structure and limited pedestrian paths. One of the fundamental design objectives was to highlight those established trees and capitalise on the shade and structure they provided. The design achieves this with the removal of unnecessary clutter, opening up the space. Two of the largest figs are the main focus of the park, highlighting the new playground and amenities hub.

HASSELL created the master plan, consulted the community and provided documentation and construction advice for the park and new buildings. The design improves the quality and scope of the park for both the local community and visitors by upgrading the facilities for passive and active recreational users, as well as catering for families by introducing a larger district playground and new amenities hub. The improvements included the building structures, paving and paths, sporting facilities, lighting, furniture, planting and playground. The design of Turruwul Park is an excellent example of how the existing qualities of a site can be identified and sensitively enhanced by the careful insertion of new design elements to add structure and create a better functioning and visually enhanced outcome. The upgrade reinvigorates the park as a focal point for the area, a resource for active recreation, sports users and quiet relaxation. The facilities include a new playground, which incorporates a series of activity zones linked by planting, boulders and timber decks, set in the shade of a large fig tree. The parks sporting amenities have been upgraded with improved sports pitches, tennis courts, hit-up wall, cricket nets and basketball court.

A series of new paths, including a tree lined avenue

around the perimeter and along key desire lines, allows pedestrians greater interaction within the space and the park facilities. Challenges encountered during the project led to innovative construction solutions for pavements to protect tree root zones. Pathways were lifted over the effected root zones by the introduction of concrete bridging structures supported on strategically positioned piers that avoided all major roots and allowed a concrete boardwalk to be supported without any compaction over the root system. The construction of the boardwalk also allowed air to circulate over the root zone and water to permeate, therefore providing a path network that did not adversely impact the existing trees. HASSELL worked closely with the City of Sydney arborist to produce a design that was sensitive to the existing trees and coordinated tree protection throughout the construction process.

The new Turruwul Park kiosk and change rooms provide the community with new facilities for their sporting and leisure activities. The public buildings, placed delicately either side of a venerable fig tree, serve as a focal point for the upgraded park. A key feature of this project is the timber facade. A woven rhythm of blackbutt battens forms a screen. The zigzag cut battens create an intricate pattern. The outline was designed to enable two adjacent battens to be cut from one piece of timber, minimising waste. A simple change to the cutting schedule enhanced the mottled and textured character of the timber, conveying the sense of the project's craft. The buildings provide a highly textured setting for community activities, achieved with an ultimate simplicity through the repetitive patterning within a simple form.

The client supported many ESD initiatives including the use of photovoltaic cells, solar hot water heating, natural ventilation and lighting, rainwater reuse and the use of sustainable materials. The project uses bore water to irrigate all soft landscape areas. Run-off water from the paths and tennis courts is treated via grass swales before being released into the stormwater system.

维多利亚公园

Victoria Park

撰文：Sharon Wright Deborah Eastment Tony McCormick   图片提供：Alexandra Shimo   翻译：刘丹春

总平面图

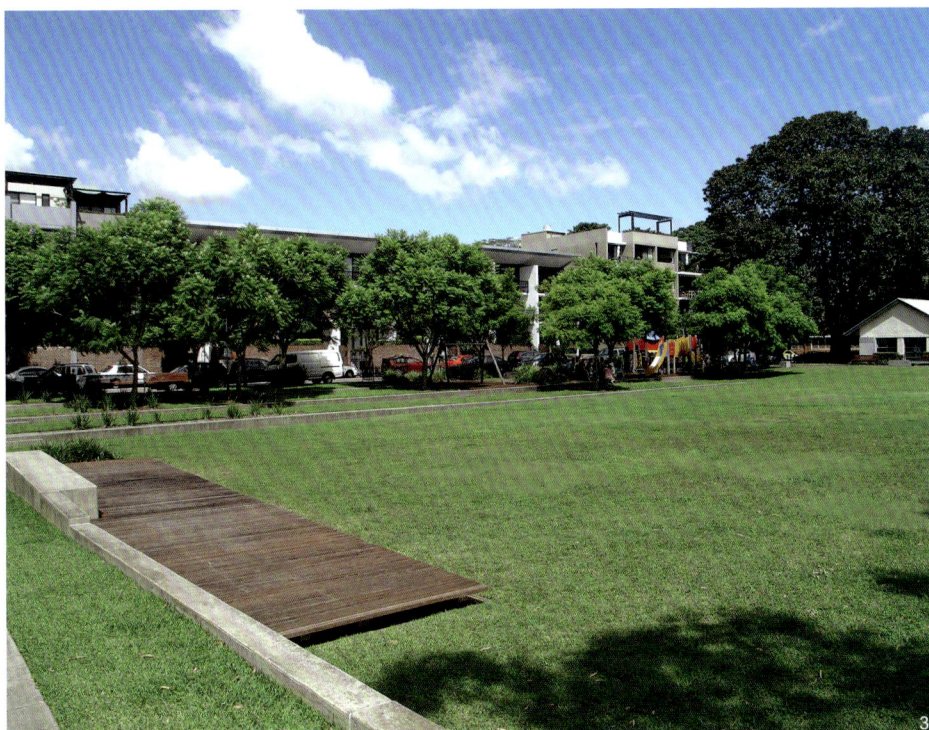

图例

小巷自行车道
种植有大型树木的住宅区道路
带有中央隔离带的街道
种植有常绿植物的商业区道路
不带中央隔离带的街道

1   木地植物景观 1
2   Joynton 公园内的野餐区与烧烤区 1
3   木板遮挡的蓄水泵
4   Woolwash 公园内的湿地景观

该项目位于澳大利亚悉尼市，占地 240 000m²，是一处集中高密度住宅、多种商业和零售设施于一体的多功能开发区，可容纳 5000 人。

该地区曾是辽阔的湿地和泻湖系统的一部分，从百年纪念公园延伸至植物湾，绵延 10km，被称做植物学湿地。1789 年以前它是"世界上最丰美的草地"之一。由于两个世纪以来的城市化压力，该场地发生了巨大的变化：这里曾先后建立赛马场、汽车制造工厂和军工厂，场地复杂的生态系统遭到破坏；从赛马场赛道建成开始，场地的生物群落价值也遭到严重的破坏。

设计师在这块棕地上应用了景观的都市主义原则，将复杂性再次引入自然、社会和文化环境中来，有益于新社区的发展。新南威尔士州政府将该项目作为典范，因为它展示了如何将滨水城市设计和文化设计原则相结合，建成较高密度的城市内部开发区。

Joynton 公园规划图

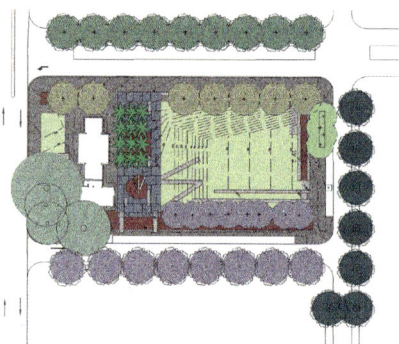

Tote 公园规划图

1999 年，在一项关于开发该区域的总体规划竞赛之后，HASSELL 公司和新南威尔士州公共工程部共同设计出总体改良计划和城市设计指南，决定了该地区的布局和建筑形式。进行布局和建筑形式设计时需要考虑该地区的洪水因素，以及该项目毗邻悉尼国际机场的飞行航道等因素。

该景观设计理念体现了区域范围内的水资源管理战略、新开发区和自然湿地系统中水的处理方式、区域连续性以及社区发展这四个与时间和地点相关的关键问题。

街道布局通过贯穿整个区域的网格保证了道路的通达性，东西走向的街道的中央隔离带是生态洼地，可以收集雨水径流并进行初步的净化处理，南北走向的街道则为传统布局。洼地结构与优雅的桥梁结构、本地物种和引进物种形成的巨大的"调色板"相结合，整体和细节都十分完美，使人行道景观更为赏心悦目。

该项目凭借其创新的生态可持续水系统超越了人们的期望。洼地渗滤系统调节该区域公路的首次冲洗水的质量，过滤后的水经过再循环应用在 Joynton 公园的著名水景中。植物的选择和生物栖息地的创造将

自然进程重新引入，并提高了生物多样性。本地物种主要用在街道和公园中，以建立该区域特有的群落并将维护需求降到最低。高品质的、密集的城市环境中融合了这种可持续的生态设计，使该项目成为全澳大利亚都市背景下滨水城市设计的基准。

该项目具有丰富的空间形式和材料组合形式，设计师选用本地湿生植物，并调整地貌以满足水管理的水量需要。社区具有休闲功能，三个主要公园满足了新社区的多重需要。

托特公园阐释了该区域的历史，着重强调了罗斯伯里马场及该马场的开创性作用。为了满足儿童的喜好，设计师还在这里设置了适当规模的游乐设施。

Joynton 公园拥有巨大的空间，有正规的烧烤和野餐设施、亭式建筑，并能够提供夜间主题照明，可以满足非正式团队运动、大型家庭聚会和各类社区活动的需要。

纳菲尔德公园拥有各类运动设施，满足了青少年的需要。公园由公路环绕，公路另一侧为住宅，有利于被动监测。

显而易见，公园和公共区域的设置成功地迎合了

社区和住户的需求。以前在利兰汽车制造厂工作的员工现在也会定期来到这里聚会，展示他们生产的汽车。

该项目证明了自然重建进程或"生态建设"可以在城市环境中进行。以前，自然系统在支撑景观、促进人与自然交流以及建设自然进程方面的价值，在郊区的土地开发中得以充分展现，但在城市里却被认为是行不通的。维多利亚公园重新定义了自然系统的作用，并揭示出"景观即是装饰"这一观点的不足之处。

该项目的公共区域，不仅为社区提供了宝贵的休闲资源，还重建了一度遭到破坏的自然进程并且丰富了生物群落的多样性。

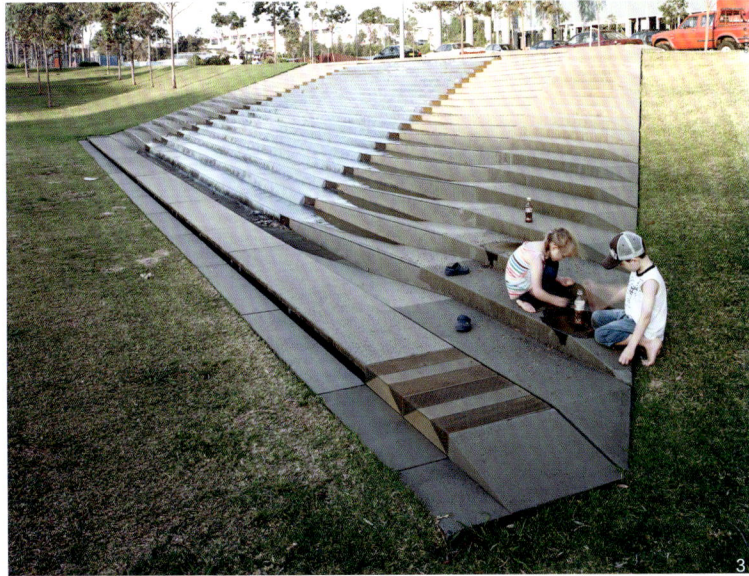

1　儿童游乐场

2　Joynton 公园内的野餐区与烧烤区 2

3　孩子们在公园内玩耍

4　独具特色的水景

5　Tote 公园细部

Victoria Park is located in Sydney, Australia. It is a 24-hectare mixed-use development, with medium and high-density housing, commercial and retail facilities for a population of 5,000.

The site was once a part of a large wetland and lagoon system extending over 10km from Centennial Park to Botany Bay, known as the Botany Swamp. In 1789 it was "one of the finest meadows in the world". In response to two centuries of urban development pressure the site was changed dramatically. Its complex ecosystems were obliterated first by a racecourse, then a car manufacturing plant, then a military store. Each change led to greater simplification of natural processes and from the time of the racecourse track, greater erosion of the community value.

In the simple form left by the military, this brownfield site gave the design team the opportunity to apply the principles of landscape urbanism by reintroducing complexity into the natural, social and cultural environment for the benefit of the new community. The result is a project the NSW State Government uses as an exemplar of how to integrate water sensitive urban design (WSUD) and cultural principles into higher density inner city developments.

In 1999, HASSELL and NSW Public Works were engaged to prepare the Refined Master Plan and Urban Design Guidelines following a master planning competition for development of the site. This determined the future site layout and built form. These were heavily influenced by flood behaviour in the area and the location of the site relative to the Sydney International Airport flight approaches.

The landscape design concept embodies four key principles that relate to its time and place: site-wide water management strategy; interpretation of water in the new development and in the natural wetland systems; site connectivity; and community development.

The street layout provides simple and legible connectivity via a grid through the site with the east-west streets containing bio-retention swales in medians to collect runoff and perform initial pollution trapping. North-south streets are more traditional avenues. Utilitarian structures within the swales are combined with elegant bridge structures and a strong palette of native and endemic species to deliver a wonderful sense of enclosure and detail, greatly appreciated at the pedestrian pace.

The project exceeded expectations by virtue of its innovative and ecologically sustainable water management system. The bio-retention swale infiltration system regulates the quality of first flush water from the site's public roads. The filtered water is intercepted, recycled and exposed to view at the site's notable water features at Joynton Park. Plant selection and habitat creation reintroduce natural processes and promote biodiversity. Native species are predominantly used in streets and parks to re-establish endemic communities on the site and minimise maintenance requirements. That this has been seamlessly integrated into a high-quality and dense urban environment provides a benchmark for water sensitive urban design in an urban context throughout Australia.

The parks in the development have richness in spatial form and materials, unified by the common thread of indigenous planting of wetland species, and a landform that is moulded to accommodate water management quantity requirements. They provide a variety of exceptional settings for the enjoyment of the new community. The three main parks have been designed to create different settings to cater for the varied needs of the new community.

Tote Park interprets the history of the site with particular emphasis on the Rosebery Racecourse and its seminal role in the invention of the Totaliser, which is now used on every race track in the world. It also caters for toddlers and young children with play facilities in a setting with an appropriate scale.

The largest space, Joynton Park, caters for informal team sports, large family gatherings and community scale events. It has formal barbecue and picnic facilities, a kiosk building that can be activated at times of demand and themed lighting to provide for evening use.

Nuffield Park has robust surfaces and "balls and wheels" facilities for the demanding use of teenagers. All parks are surrounded by roads fronted with housing to promote passive surveillance.

The success of the parks and public domain in catering for the specific needs of various community and resident groups is obvious. Right down to the former workers at the Leyland car manufacturing plant that now regularly use the site for reunion functions and concourses showcasing the cars they built at the site.

Victoria Park is proof that reconstructed natural processes, or "constructed ecologies" can work in urban environments.

The value of natural systems to sustain the landscape and expose people to nature and its processes has often been practised in greenfield development in the suburbs, but deemed unworkable in urban situations. Victoria Park redefines the role of natural systems, and exposes the shortcomings in the ideas of landscape as ornament.

Its public realm not only provides a valuable recreation resource for the community, it also sets about reconstructing the natural processes and community richness that once prevailed.

昔日铁路沿线，今日中央公园 —— 门多萨公园

Railways in the past, Central Park in the Present — Mendoza Park

撰文 / 图片提供：Jimena Martignoni　　翻译：刘宏阳

门多萨市位于阿根廷安第斯山脉脚下，在行政划分上隶属门多萨省。该地区属于典型的地中海气候，其葡萄酒酿造工业具有得天独厚的优势。虽然门多萨市的气候与沙漠气候相似，但这座城市的绿化率很高，随着历史变迁，慢慢拥有了一种独特的城市魅力。这座城市的地标是城市沟渠系统，远处的河流为沟渠提供水源，贯穿整座城市，从而形成一个卓有成效的灌溉系统。

门多萨市的总人口为15万，有一座大型公园和一系列小型公园和广场。然而，在20世纪90年代，却出现了城市居民向城市外围迁移的现象。为了改善这一状况，2000年，当地市政府决定在城市中心一块废弃的土地上兴建一座大型公园，以有效改善门多萨市的人居环境。

该场地占地面积约14万平方米，原本属于阿根廷国家政府，曾是重要的铁路服务区，建有一座火车站和许多铁路仓库。经过协商，阿根廷国家政府将场地的土地所有权划归门多萨市政府所有。门多萨市政府举办了一场国际招标大赛，在参赛作品中择优选择出改造该场地的设计方案。项目的建设分为两期，一期工程主要是建设占地90 000m²的公园主体，二期工程是在修缮后的铁路仓库中建设文化设施。

在门多萨市政府举办的这场国际性竞赛中，来自布宜诺斯艾利斯的一个建筑师团队最终夺得大奖，他

们往返于布宜诺斯艾利斯与门多萨之间，反复考察场地，最终确定了设计方案。目前，门多萨公园的一期建设已经完工，修缮铁路仓库的二期工程于 2010 年施工。

该场地原为铁路枢纽区域，建有大量的铁道，设计师在进行设计时充分考虑了场地的现状。该项目的主设计师、建筑师 Daniel Becker 说："该项目的设计并不是基于场地的地理位置（场地恰好位于门多萨市的中心），而是基于场地原有铁路线的几何构造。"

场地原有的铁路线纵横交错，形成多种多样的几何图形，设计师巧妙地利用这一现状，将它们视为具有特别意义的"疤痕"。这些"疤痕"充满象征意义，

构成了该设计的基础。一方面，场地原有的西北－东南方向的铁路仓库决定了新建的步道和公园其他建筑元素（如地面铺装、绿化区和人工湖等）的走向；另一方面，场地区域的划分在轮廓上与原有的铁路线相呼应。

门多萨公园的主体部分占地 90 000m²，而原有的铁路仓库占地约 5 万平方米，二者由场地原有的一条南北向的机动车道分隔开来。在市政府组织的国际性竞赛中，设计要求之一便是在公园主体与铁路仓库之间建立一种有形的联系。为此，设计师们在二者之间新建了两座人行桥，桥下建有地下通道，以保证这条机动车道的正常通车，并且在公园边界建造了挡土墙。

新建的人行桥与场地原有的铁路线相互呼应，直接连通着公园的带状步道。

该设计的另一大突出贡献是使门多萨公园突破了大型公园的功能限制，具备城市广场的功能，成为大型公园与城市广场的功能结合体。设计师们在场地内设计了两种截然不同又清晰可辨的景观元素：一是由草坪斜坡和树丛组成的绿色"缓冲"带，直观地将人工湖围合起来，并且延伸到公园周边；二是公园中心的传统形式的广场，与其说这是一座大广场，不如说这是由一系列彼此相连的小广场组成的广场区，适合举行各种各样的活动。为了鼓励人们经常到此进行体育活动，设计师们还设计了一条环形慢跑道，与其他

道路位于同一水平高度上，跑道两侧树木林立、花香浓郁。

公园内的 3 条二级道路勾勒出公园的轮廓，沿着这 3 条二级道路分布着数个停车场，恰当地解决了设计大赛的要求之一——停车问题。

该项目的中心部分由一系列水平面组成，大部分水平面都交替排列着绿色步道、木质步道和水台等。考虑到场地的景观元素多是西北－东南走向，设计师们新建了一条长长的混凝土挡土墙以平衡场地的高差，挡土墙本身在空间上界定了各个平面，同时也是水景的重要构成元素。设计师将场地的高处建了露台，俯视着公园的低处空间，而高处与低处通过不同宽度、不同形状的台阶相互连接。

轮廓分明的带状元素清晰地标示出公园的走向，并且界定了公园内的各个区域；在门多萨公园的东南端，挡土墙将公园与时钟广场（Plaza del Reloj）连接起来。时钟广场是一座小型广场，主要作为行人通道，

且设有一座日晷。门多萨公园内建有许多水景，如瀑布和喷泉等，水的灵动与清凉使公园里的游人络绎不绝（在酷热的天气里尤为如此），而在夜晚，通过精巧的照明设计，水景散发出别样的魅力，成为游人集会的主要场所。

公园内建有一座大型人工湖，灵动的水是人工湖景观最为重要的元素。人工湖两侧是铺满绿草的斜坡，游人或漫步在草地上，或在树下休息乘凉，与大自然亲密地接触。青青的垂柳掩映着木质平台，鸭子们在湖里游来游去，远处青山在望，与公园内的地面铺装（如石子路等）形成视觉反差。一座外形酷似码头的平台延展到人工湖的上方，这座平台是孩子们的最爱，他们从山坡上一路奔跑着来到这里，在大自然的怀抱中畅快淋漓地玩耍。在这座平台的尽头，场地原有的一棵大树被完好地保留下来，为游人提供荫蔽的同时，见证着场地的旧貌与新颜。

设计师们同样关注细节的设计，包括每一处景观

小品、灯柱和游乐场等。尽管设计师们设计的景观小品没有全部应用在场地中，但是选用的景观小品均为公园增添了一种建筑魅力。公园内的公厕由混凝土建成，集中在一个方形区域内；公厕朝向公园的主外立面由木质长板组成，形成一个大的木质平面，与公园内的带状线条相互呼应。公园内的售货亭和保安室的面积比公厕的面积小很多，但是采用的设计手法相同，均采用混凝土与木质长板相结合的建筑形式。

公园内的游乐场位于挡土墙和瀑布前方，游乐场内各种设施齐备，分布错落有致，与公园内的许多设计元素相互呼应，使游人不禁想起场地原有的纵横交错的铁道。公园的植物景观遵循"诗意"的设计原则，一排排繁茂的花丛穿插在游乐场中，高大的白杨树成排林立在人行桥两侧，蓝花楹点缀在小路两旁。在公园一期建成后的短短几年之中，公园内的游人络绎不绝，熙熙攘攘，分外热闹。然而公园的维护成本却高于预算，设计师们不得不放弃了部分最初的设计方案。

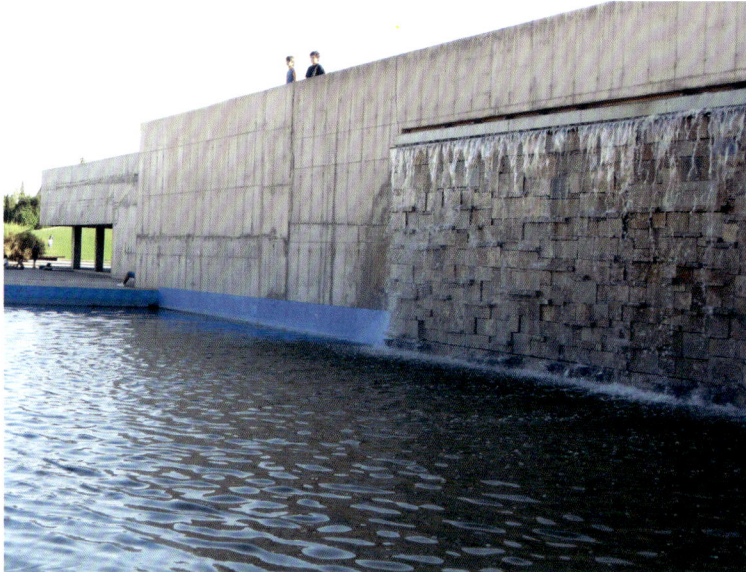

The city of Mendoza lies at the foot of the Andes Mountains and belongs to the province of the same name, in Argentina. This narrow region enjoys a typical Mediterranean climate which makes the land ideal for wine production. However, the city itself is built within a very arid and desert-like larger area which, over time, has been transformed into a green and attractive place. The construction of a system of "acequias" (or urban ditches) which are fed by distant rivers and appear throughout the city, has helped to concrete a very efficient irrigation plan and has also became the landmark of this place.

The city, which has 150,000 inhabitants, offers a large park and a series of smaller parks and plazas, but this seems to have been not enough to avoid migration from the city to the periphery during the 1990s. For this reason, the local government decided, in 2000, to create a new large park out of the conversion of a piece of abandoned land located at the exact geographical center of the city.

Covering almost 14 hectares, this land was owned by the national government and had made part of an important service area of the railway in past decades, including a main station and a series of warehouses. After some negotiations the land's ownership was ceded to the city and the local government held an international competition for the design and transformation of the abandoned area. The first stage would focus on nine hectares that would make the park itself, and the rest would provide cultural installations in the renovated warehouses.

The competition was awarded to a team of architects based in Buenos Aires, who traveled back and forth to visit the site and accomplish the design. So far, the only portion that has been finished is the park, and the warehouses renovation was started this year.

The conceptual layout of the park responds to the formal presence and significance of the railroad, or what has left of it on the area. According to architect Daniel Becker, one of the lead designers, "The project is not rooted into the geometry of the city, but into the geometry of the remains of the railways."

In this manner, the shapes and lines that crossed the land are taken as meaningful silhouettes or "scars", which become the symbol and foundation of the project. On the one hand, the northwest-southeast position of the warehouses on the site determines the position of all new esplanades and main built elements of the park, such as paved surfaces, planted areas and the lake; on the other hand, the virtual overlapping of horizontal planes and lines in the park reminds those of the old railroad.

The nine hectares that constitute the park and the five hectares that are occupied by the warehouses were previously separated by a vehicular street, running north-south. In the competition's program, one of the requests was the creation of a physical connection—specifically a pedestrian bridge—between these two areas, and the team decided to generate an underpass in order to preserve the existing vehicular flow, and to build retaining walls on the border of the park. The new two pedestrian bridges are parallel to the old industrial constructions, following the directions of their architectural lines on the land, and connect directly to the linear esplanades of the park.

Another main decision for the conceptual layout of the project was the comprehension of the park as one which is halfway between the scale of a large park and the scale of an urban plaza. As a result, two different and easily recognizable situations are generated at the site: a green "buffer", composed of lawn slopes and clusters of large trees, which extends around the entire perimeter of the park and more radically around the lake; and a central institutional-looking plaza, or series of interconnected plazas, which offer passive and active recreational areas. In order to incorporate sportive activities through which people can relate to the place in a daily basis, the project provides a jogging circuit developed at street level and framed by flowering trees.

Along three of the secondary streets that outline the park, appear also the parking lots requested in the program.

The "heart" of the park is a series of horizontal planes, mostly paved, which alternate with water surfaces, green esplanades and some wooden decks. Following the northwest-southeast direction of these compositions, the project negotiates the existing elevation changes by building a long retaining concrete wall that acts as a solid limit between planes and is also a waterfall. The upper areas of the park become terraces that overlook the lower parts of the park, and which are connected through stairs of different widths and shapes.

This strongly linear element provides a markedly sense of direction in the park and demarcate areas; in the southeast tip of the park, this wall establishes a connection with Plaza del Reloj (Clock Plaza), a small access plaza where lies a sun clock. The refreshing presence of water, in the waterfall and in the fountain underneath, attracts groups of people during hot summer days and, at night, when dramatically illuminated, turns into a place of congregation.

Additionally, water comes out as a major presence when defines a large artificial lake. Framed, at two of its sides, by green slopes where people gather and sit around underneath the trees, this aquatic component provides a natural ambiance within the park. Wooden decks whose edges are planted with willows, ducks swimming, and the distant view of the mountains, contrast with the paved and stone surfaces of the adjacent areas. Jutting into the water, a pier-like platform offers an arrival area for visitors sitting and walking on the slopes, especially kids who run downwards and want to get as close as possible to the water. At the very end of this pier, an existing large native tree was preserved as a witness of the process of transformation, providing shade and shelter.

As part of the project, the architects designed every single piece of furniture, the lighting posts, and the playgrounds. Although not all of the furniture pieces were actually built and incorporated into the site, there are some elements which add a strong architectural image. The public restrooms are concentrated in a box-like volume made of concrete; its main façade, the one facing the park, is finished with wooden pieces that, together, appear as a single plane which repeats the linear silhouettes and constructions in the park. With identical design, but responding to a much smaller scale, are the kiosks and security booths.

The different modules that compose the playgrounds are located, one after the other, in front of the dividing wall and waterfall, confirming a subtle dialogue between the different elements of the park and the reference to the horizontality of the lines of the past. The planting plan follows this poetic too, and creates some rows of dense flowering shrubs along the playgrounds, or rows of large poplars and jacarandas along the connecting paths and pedestrian bridges. After a few years of the completion of a first stage, this park looks quite settled and it's usually crowded. However, the level of maintenance is not the expected and part of the original design has decayed.

横滨太阳城公园

SunCity Park Yokohama

撰文 / 图片提供：SWA 集团　　翻译：张晶

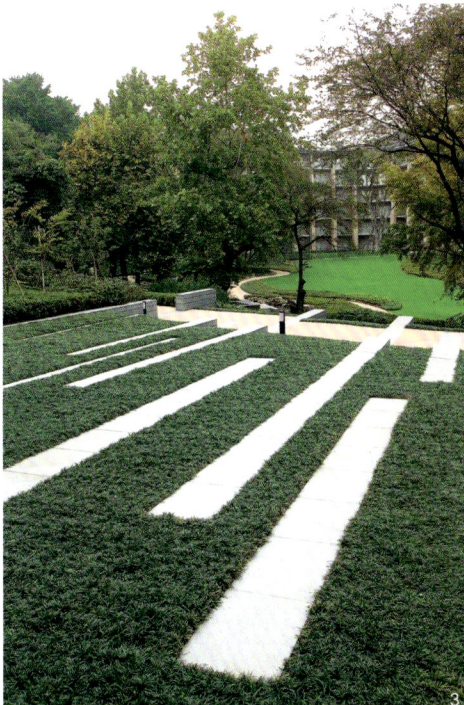

1 环绕太阳城的林地
2 场地原有的植被
3 独特的地面铺装与植被
4 水景与银杏树

总平面图

在日本，65 岁以上的人口占全国总人口的 20%，而老年人对高档生活社区的需求也越来越高。日本的老年人同美国的退休人员不同，他们比较喜欢在提供康乐服务和娱乐消遣活动的景观式酒店公寓独居。半个多世纪以来，开发商逐渐意识到打造一处带有花园和公共休闲场地的温馨居住环境比修建慈善机构更为重要。

保土谷区的地势高于横滨市区，地理位置特殊，这里的山顶上曾建有一个炸药厂，从那里可以俯瞰横滨港全貌。为避免发生意外火灾事故，这里一直人流稀少、发展缓慢。随着火药生产方法的改进，炸药厂迁往别处，原工厂的土地被认定为公共财产，根据城市规划要求，这里只能用于开发公共和私人的健康福利设施。

该项目所在场地自然风光优美、三面开阔，在全日本是数一数二的老年人生活社区。公园高居山顶，林地环绕，其人性化的建筑布局和卓越的景观设计，使之成为日本老年人生活社区建设史上的里程碑。

在工程建设之初，设计师们对项目地进行了可行性研究，并对一位竞标开发商提出的建筑方案进行了评估。可行性研究报告表明，相对于修建不规则的阶梯式错层建筑方案，在同一水平面修建楼层相同的 1

栋～2 栋大楼的方案更加可行，这样既可以满足项目要求的单元数，又避免了阶梯式多层建筑施工的不便。在特聘建筑师的协助下，设计方案最终出台——修建 2 栋 6 层的楼房，包括 480 个单元房和一个有着 120 个床位的康复中心，并且保留此地独特的自然风光。

该项目面临的挑战是既要保护自然资源，又要开辟出可供老年人全年进行活动的各种户外场地，并设计一个可以完好保留花园原有树木的景观，这些树木一直受到土地主人和社区的珍视。施工分阶段进行，既保留了原有的树木，也实现了土石挖填过程中资源的平衡利用，不需要将土石运出工地。

设计规划包括两栋由一个封闭式小桥相连接的独立"村屋"式建筑，看起来处于一个天然的山谷之中。这两栋"村屋"两侧的楼房可用于居住，同时也是公园景观的一部分，从房间里可以欣赏到花园和林地，向东又可以远眺横滨市。

两栋"村屋"和康复中心北面保存完好的参天大树勾勒出了"漫步花园"的边缘线。无论是从"村屋"里的公共大活动室或者公寓房的房间向下望，还是从连接"村屋"的封闭式小桥上看过去，都可将"漫步花园"的美景尽收眼底——一个宽阔的大草坪、一个茂密的森林花园和一条潺潺的小溪。从大楼门前的石

阶信步穿过草坪或森林花园中的石子路，可以来到一个活动区。活动区侧面有一个水畔凉亭，入口是带有屋顶的大门，极具日式传统风格。这个大门是从过去的一幢别墅中保留下来的，森林花园和小溪经过别墅延伸到石桥处的汽车道。

大楼两侧的空地，以及大楼和公园边界之间的一些间隙都被修建成各具特色的花园。其中一些花园只对私人开放，提供单独休养的空间，其他人不允许进入，只能从房间里或其他公共花园远远望一望。

横滨太阳城公园的保留景观和新造景观以其突出的建筑设计和内部装潢设计赢得了赞誉，亦给每位曾经体验这个老年人生活社区的人留下了深刻印象。

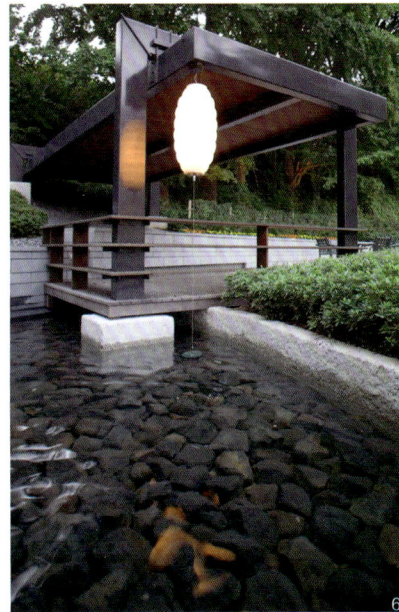

1　公共步道
2　池塘
3、7　石阶小桥
4　水景
5　石质喷泉
6　日式的传统灯笼

With twenty percent of Japan's population at age 65 years or older, the demands for high quality residential communities for seniors are growing rapidly. Not unlike American retirees, many of Japan's seniors are looking for independent living communities with hotel-like amenities, activities, and lush grounds. It has been important to Half Century More, the developer, to create a caring, respectful environment with gardens and public spaces that feel residential rather than institutional.

The unique character of the Hodogaya site above Yokohama City, a former explosives manufacturing facility on a hilltop overlooking the port city of Yokohama, came from its own isolation from other development and population, a safety precaution to prevent catastrophic fires that may have resulted from accidents. As development slowly encroached and production methodologies changed, the explosives manufacturing moved elsewhere and only company offices remained. The land was recognized as an asset to the community at large, and zoning restrictions were placed on the parcel to limit development to public and private health and welfare facilities.

The natural beauty of the site with preserved open space on three sides, makes SunCity Park Yokohama one of the most unique senior residential communities in all of Japan. The hilltop setting, woodland edges, and sensitive building layout, in conjunction with the introduced landscape present SunCity Park Yokohama as a landmark senior community.

The landscape architects' role began with a feasibility study of the site and a review of the existing development scheme proposed by a competing senior housing developer. The feasibility study confirmed that the previously proposed and approved unit count could be achieved with a single or two building project on level building pads versus a sprawling multilevel hillside development that did not fit the operational model of the interested developer. Together with the selected design architect, a site plan sensitive to the natural character and charm of this very unique site was proposed, resulting in two six-story buildings with a total of 480 Independent Living units and a 120-bed Care Center.

The challenge was to preserve natural assets and provide a variety of outdoor spaces that could be used throughout the year by seniors, and to design a landscape that would preserve the garden elements that were treasured by the landowner and the community. Grading was perhaps the most important effort to protect and preserve existing trees and to balance cut and fill as no material could be taken off site.

The site plan includes two single building "villages"

connected by an enclosed bridge spanning a natural valley. Residential wings of the two villages extend into the landscape offering views into gardens, woodland preserve, and stunning distant views to Yokohama City to the east.

The Villages, and the carefully preserved mature trees north of the Care Center, frame the Stroll Garden. The Stroll Garden, viewed from the main public rooms of both Villages and residential units above, and from the enclosed bridge connecting the Villages, features a large open lawn and woodland understory garden separated by a gently cascading stream. Stroll paths from stone paved terraces immediate to the buildings lead through the lawn or woodland garden areas to an activity terrace flanked by a waterside pavilion and a traditional styled Japanese roofed gate that could be a remnant from a villa of an era gone by the woodland understory garden and cascading stream continue out to the stone bridge of the entry drive.

Various other gardens occur around the site created by the various wings of the two buildings and the many spaces between the buildings and parcel edge. Some of these gardens offer private retreat and respite while others are solely for viewing from residential and public spaces alike.

The preserved and created landscapes of SunCity Park Yokohama compliment outstanding architectural and interior design, and make it a memorable senior community for all who experience it.

1 池塘边的小亭子
2 步道
3 花园入口

安阳之巅

Anyang Peak

撰文 / 图片提供：MVRDV　　翻译：谷晓瑞

这座主题公园位于韩国安阳市旁的群山中，其巧夺天工的设计成为了一道亮丽的风景。

上世纪七八十年代，安阳公园曾是极受人们欢迎的旅游胜地，但由于景区的发展落后，近十年来渐渐地被人们遗忘了。曾经的特色室外游泳池也失去了往日的吸引力，曾经独具特色的园林景观现在也只是个山谷入口，周边零散地分布着几家餐馆和酒吧。而且如今各地景观比比皆是，发达的交通也使人们可以去远处欣赏更优美的风景。

1999 年，安阳公共艺术计划的发起给这里带来了重生，它将人文与艺术完美地结合在一起。来自世界各地的建筑师、艺术家和设计师们会聚一堂，共同设计建造了包括会展厅、展览馆和街道设施等在内的 50

项永久性建筑，以及一些临时设施。

让这里重现生机的方法就是充分利用自然景观，着重挖掘其自然本色。设计师提出了一个创意，在山顶修建一座观景塔，通往山顶的小路便成了该主题公园的亮点。蜿蜒的小路组成了塔楼，重新勾勒整座山峰的外形。

山顶上的两条轮廓线限制了小路的外形——内外各一条螺旋线。两条轮廓线蜿蜒而下，小路的宽度也随之变化。最窄处的宽度为1.5m，这也是两条轮廓线的基准宽度。长为146m的小路构成4个圆圈，坡度

约为1/10，形成一个垂直高度为14.6m、面积约为160平方米的"山峰"。

塔楼的主体采用钢铁结构，小路地基采用垂直的钢管支撑，地基之上是定制的铁板，上面又覆盖了一层木板。

这一新的建筑物不仅构成一个观景塔，同时也是一件艺术品和一个表演空间。其内部空间可作为展览馆，用于举行小型的展览会或作品展。如果在山脚搭建表演舞台的话，观景塔则是最佳的看台。这条环绕山顶的小路将山巅变成了一处富有趣味的景观。

In the mountains next to Anyang, a city in South-Korea, a theme park is developed that stresses the beauty of the site through a series of artistic interventions.

Anyang resort is a park which boomed during the 70-80s and been redundant during the past decade because it couldn't follow up the change of society. The outdoor swimming pool which was the most important program couldn't attract people anymore because they could easily find the better one. And as the traffic improves, they could travel further away for nicer scenery. Even though it has a spectacular landscape, it just functions as a mountain entrance with scattered restaurants and bars.

In 1999, the Anyang Public Art Project (APAP) was launched that would regenerate the area by bringing people and art together. Architects, artists and designers from all over the world were invited to create a total of 50 permanent works, such as an exhibition hall, pavilions and street furniture, and several temporary installations.

One way to revitalize this area is to mobilize its natural wonders, intensifying nature. MVRDV proposed a viewing tower supercharging the hill into Anyang peak. The path leading up the hill, an essential element of the park, is used as a tool to generate this idea. The spiral path becomes the tower, extending the hill seamlessly and reshapes the peak.

Two contour lines from the top were used to shape the path. One forms the outer spiral and another one forms the inside spiral line. As these two contours offset inwards, the width of the path varies. The minimal width, which is 1.5m, was the guiding line for these two contours. And the inclination of the path was fixed as 1/10 slope. A 146m long path with four rings forms the 14.6m height peak (92m+14.6m=106.6m) which covers an area of 160m^2.

The tower's structure is made of steel. The substructure for the path is suspended off the vertical steel poles. Customised steel plates are fixed onto the substructure and covered with wooden planks.

The new structure functions not only as a viewing tower alone, but also invites people to experience the tower as a piece of art and performance space. The internal void acts as a pavilion; it can hold a small exhibition or installation. The viewing tower becomes a tribune when performances are staged underneath on the top of the hill. The path encircling the peak turns it into a destination.

两百周年纪念公园

Bicentennial Park

撰文：Jimena Martignoni　　图片提供：Teodoro Fernandez Office　　翻译：申为军

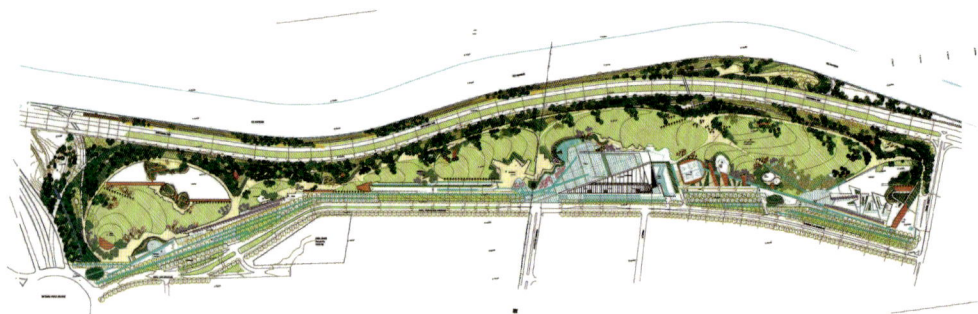

总平面图

1　公园北门入口处水池
2　广场铺装与植栽
3　通往山脚的步道 1

该项目位于智利首都圣地亚哥，占地面积为 27 万平方米；到 2008 年 3 月为止，已经建成了 12 万平方米。为建造这个公园，早在 1998 年政府就举办了一场全国性的设计竞赛，并于次年公布了竞赛结果，该项目的总体设计就来自于当年的获奖方案。但由于资金和相关的法规问题，整个工程直到 2006 年才开始动工。Vitacura 区是圣地亚哥最为繁华的地区之一，位于城市的北部；该项目的设计竞赛就是由该区政府发起的。

马波乔河是贯穿圣地亚哥的一条大河，其中有一小段流经 Vitacura 区，这条河决定了整座城市的规划格局，该项目的所在地就是马波乔河边的一块狭长地段。在圣地亚哥，很多公园都位于马波乔河的两岸，这些带状的绿色区域大多在客观上具有多重功能。因此，当地居民对公共空间的模式形成了一个总体印象——高大的树木和绿色的休闲广场依河而建，整个空间与城市系统紧密相连。但是，这些公园之间相互没有连接。

项目所在区域是马波乔河惟一一段由北向南流经城市的河段，就在圣克里斯托瓦尔山脚下。公园的西侧面向壮美的山峦，东侧则与城市网络相连。河堤高出场地地平面近 4 米，与即将增建的一条快车道平行，这段高差最初体现在一道垂直的高墙上，不仅限制了设计，还妨碍建造上山公路。这条公路于 2009 年动工，切断了公园与河道间任何实质或视觉上的联系。

此外，场地还紧邻 Vitacura 区政府的办公楼。该办公楼建于 2004 年，其设计方案源于另一项设计方案的征集竞赛，竞赛的要求之一就是要设计一个可与公园直接相连的铺装区或广场。在这个半开放广场的地下，还建有一个专供区政府工作人员使用的地下停车场。

建筑师 Fernández Larrañaga 在该项目的整体规划设计中，全面考虑了这些相对复杂的场地状况。针对现有的地下停车场，设计了一个可以遮掩地下空间的屋顶式结构。这是一块单独的铺装平台，不但与公园相通，也可为工作人员在办公楼延伸屋顶下举行半公共活动提供场地。设计师将服务于政府办公楼的广场设计成面积为 4856 ㎡ 的一系列平台，这些平台延伸向公园，空间比例十分完美。平台的边缘使用了各种几何形状作为过渡元素，如台阶、缓坡等；广场最长边的前缘处则设有一个线形的饮水器。

出于对原有场地环境的尊重，设计师设计了一个遵循圣地亚哥传统造园方式的简单的方案。为了消除场地与公路的高差，设计师在公园的西侧设计了连续的、高度不断变化的绿地，其坡度变化在 4.8m ～ 12m 之间，从而形成一个自然的绿色边界。这片绿地在视觉上与位于背景处的圣克里斯托瓦尔山融为一体，同时也将高速路上的汽车阻隔在人们的视线之外。设计师在这些人为制造的陡坡上种植了许多当地树种和高大的棕榈树，与不远处连绵起伏的山峦形成了连续的绿色，保留了圣地亚哥多山丘的城市特征。

公园东侧被定义为城市边缘区，以规则的网格结

构种植的梧桐树突出强调了道路的几何布局，这种线性布局只在特定位置被打断，如入口、斜坡、成片种植的观赏禾草。区域内的步道几乎全部采用铺装，部分地段还使用了鹅卵石；为了在形式和视觉上与之对应，设计师在路旁设置了一些低矮的石墙，形成更加自然的边界，让人联想起智利乡村特有的标志性矮墙。

一道石砌的挡土墙解决了公园东侧的高差问题，宽阔的台阶和斜坡通向公园的绿地，融入那些起伏的、色彩柔和的绿化带，围绕在办公楼广场周围。这条步道最为开放的一段位于街道旁，与那些通向公园的路段有所不同。这种"平台＋路径"的设计理念在公园中不断出现，尤其从西侧的山峦望去，远处城市中新建的高塔和低处的区政府办公楼及广场都一览无余。从对面的市政道路看向西侧地势高的小径（也是骑车道），不断起伏的山峦和斜坡上会出现很多骑车的身影，在静止的线性景观中缓缓移动。

这座典型的现代城市公园的西侧为自然景观，东侧则为城市景观。在公园中，游人在开阔的草坪上或坐或躺，或散步、慢跑、野餐、嬉戏，享受阳光，放松身心；公园里还设置了一些私密空间，并配有弯曲的混凝土长凳，步道的两侧亦有很多典型的木质长椅。这些空间周围种植了花灌木、小型树木和观赏禾草，使这一大型场地的景观富于变化。

在连接植满梧桐树的小径和休闲广场的主阶梯前，分布着一系列现代化的游乐园，其中蓝色的绳索结构总是能吸引很多孩子。这里每天都是一派热闹的景象，特别是周末的时候。

公园的北部目前是整个项目中最成熟的一处景观，位于政府办公楼后面，属于前期工程的一部分。这片水景区由一系列与地平面高度一致的喷泉水池组成，水池边缘处一道道窄窄的混凝土线条彼此纵横交错，其间点缀着纸莎草及多种观赏性强的水生植物。很多情侣喜欢来此欣赏宜人的水景，当地居民也喜欢一家人来此游玩，因此这里还种植了很多茂盛的水生植物，在水中游弋的天鹅常吸引孩子们前来喂食。场地最北端、邻近水景区有一家充满现代设计风格的餐馆，这

也是一项设计竞赛的获奖作品，设计者是著名的智利建筑师 Smiljan Radic。办公楼的露台与公园的这部分区域及另一个入口相连通。

公园整体竣工的时间配合于 2010 年举办的各种庆祝活动，以纪念智利、阿根廷、墨西哥、委内瑞拉、厄瓜多尔、玻利维亚、巴拉圭这七个拉美国家独立两百周年。事实上，公园的名字就源于此。

如今，该项目已经成为备受 Vitacura 区及附近居民喜爱的休闲场所。此外，圣地亚哥市政府还有意在远离市中心的另外一段河流沿岸再建造一座线性公园。

1　当地的棕榈树 1
2、3　鹅卵石步道
4　步道
5　水池 1

1、4 水池周围植物
2 通往山脚的步道 2
3 当地的棕榈树 2

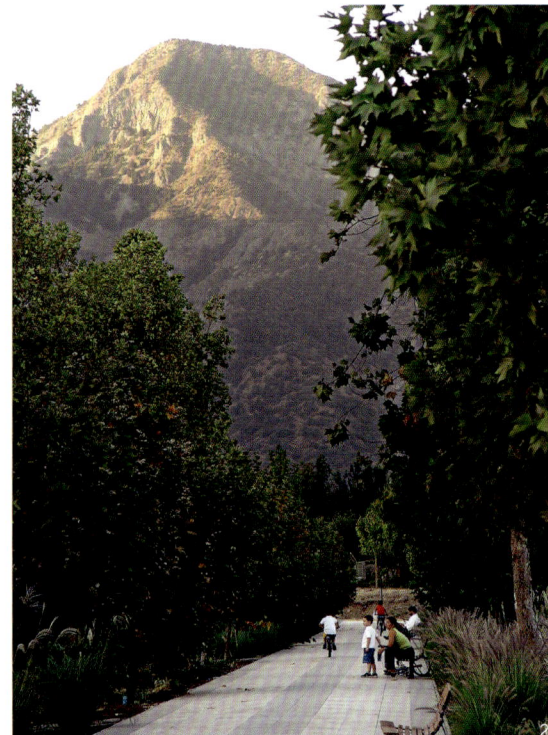

In March of 2008 were completed the first 30 acres, out of a total of 67, of Parque Bicentenario, a new linear park in Santiago, Chile's capital city. This park is the result of a national competition held in 1998 and whose results made public in 1999, but because of financial and jurisprudence issues the construction of the park wasn't started until 2006. This piece of land had historically been owned by the national government and had to be transferred to the Vitacura Municipality, one of the most prosperous municipalities of Santiago, located in the northeast area of this city, and which had proposed the contest.

The site is a strip of land which develops along a small part of the Mapocho River, the most important river and defining natural element of the city's physical structure, within the Vitacura Municipality's geographical limits. In Santiago, the most traditional public parks are those which develop along the Mapocho River, as linear green areas that offer diverse uses, mostly passive. These parks create, together, one single image in the collective imagination of locals: public spaces with large trees and green esplanades in close connection to the river and to the urban system. However, they are not always physically connected with each other.

In this case, the site extends along the only portion where this river crosses the city from north to south, at the foot of the San Cristobal hillsides; the full extension of the west side of the park facing the "cerros" or mountains and the connection with the urban grid given along its east side. Approximately 13 feet above the site's ground level and parallel to what will be a new fast-lane of an existing roadway, runs the embanked river. The elevation change between the river/road embankment and the park, formerly materialized with an upright wall, establishes a solid limit and the presence of the future uphill roadway, whose construction is planned for 2009, eliminates any real physical or visual connection between the park and the water course.

At the site, another existing condition was the new office building of the municipality, which was built in 2004 adjacent to it as the result of a different public competition, and whose request program asked for a paved area or plaza that would connect directly with the park. Built underneath what would be the complete surface of this semi-public plaza, an underground parking for the use of the municipality's staff had been also included in the architectural project.

Architect Teodoro Fernández Larrañaga, the wining designer, had to take this quite complex existing situation into account for the layout of the project. In response to the existing underground parking area, he designed what was a "rooftop that would cover" these underground spaces, a single paved plane which connects with the park and which also becomes the venue for semi-public activities to be performed underneath the building's semi-roofed areas. The plaza, which serves the municipality building, was designed as a 1.20 acre-paved series of terraces that embrace the construction and extend toward the park's surfaces, following a fine spatial balance. Its sides growing out as geometrical shapes toward the adjacent lawn areas

and defining some transitional elements such as steps, subtle slopes and a linear water fountain situated right along its frontline and longest edge.

With a very respectful approach for the rest of and most important pre-existing conditions he decided to create a simple plan that would, somehow, still follow the traditional manner in which the most emblematic parks had always been implemented in Santiago. In order to smooth the elevation change between the site's ground plane and that of the roadway, he modeled a continuous green rise whose percent slope ranges from 16 ft to almost 40 ft., all along the west side of the park or "its backside". This element provides a natural closing and green border visually linked with the background mountain of San Cristobal and also generates a protecting and isolating natural buffer that keeps passing cars out of sight from the park. The visual continuity between these artificial slopes, planted with native trees and palms, and the mountains behind provides an uninterrupted green background plane which preserves the constant presence of hillsides in this city.

The east side of the park is clearly defined as an urban edge; planted with a regular grid of sycamores that accentuates the geometrical layout of the path, this linear element is only interrupted, at certain meaningful spots, with entrances and ramps and homogeneous groups of framing ornamental grasses. This path is paved in almost its entire area and, in some portions, finished with cobblestone. In order to formally and visually accompany these portions, the designer added some short rocky walls which define a more natural edge and are reminiscent of the typical retaining and marking walls of Chilean agricultural valleys.

A stone-clad retaining wall solves the elevation change in this part of the site and the connection with the park's ground level is given by some wide stairs and ramps that combine with the soft green undulating surfaces that

enclose the administrative building's plaza. The most public part of this path, which develops along the street, is differentiated from those in intimate relation with the park's area with a series of steps. This concept of terraces and watching promenades is repeated throughout the park, especially from the west hillsides from where the sight of the city's new tallest towers in the distance and the municipality building and plaza below can be fully appreciated. The west upper path is also a bicycle circuit; looked at from the opposite urban path the continuous plane of the slopes and the mountains seems only broken by the sight of many bicycles moving slowly across a static linear landscape.

The space that extends between the west side or natural edge and the east side or urban edge, was laid out as a classic park within a modern city: an open meadow where to saunter, sit around, lay down, jog, picnic and play, sunbathe and relax. This esplanade-like surface offers some intimate spots furnished with concrete benches of irregular winding shapes and some other typical wooded benches placed along the pedestrian paths in the park. Framed with flowering shrubs, small trees and grasses, these places mean a change within the large scale of the site.

Right in front of the main stairs that connect the sycamores' path with the esplanade, are located a series of modern playgrounds whose blue rope structures are usually crowded with kids; a colorful dynamic scene that repeats every day, especially during the weekends.

In the north tip of the park appears what is presently the most settled composition of the project; built right after the plaza, as part of one of the initial phases, this is a series of water ground-level fountains crisscrossed with narrow concrete lines and framed with strips of papyrus, grasses and other colorful water loving species. Usually peacefully observed and visited by families and couples, this charming spot is completed with thick green carpets of aquatic plants and swans which kids love to feed. Adjoining these water fountains, at the furthermost north piece of the site, lies a restaurant whose modern design was also the result of a special competition won by Smiljan Radic, a renowned Chilean architect. The terraces of the building connect with athis part of the park and also with a secondary access.

The completion of this park coincided with the many cultural events that had taken place in 2010 to celebrate the independence from Spain in seven Latin American countries: Chile, Argentina, Mexico, Venezuela, Ecuador, Bolivia and Paraguay. In fact, the name of the park is a direct reference to the Bicentennial celebration.

So far, the park has proven to be a successful recreational place for the people of Vitacura and its neighboring municipalities and, on the other hand, Santiago now offers another linear park along the river but located in a less central part of the city.

肯沃西公园

Kenworthy Park

撰文 / 图片提供：PMA 公司　　翻译：谷晓瑞

该项目所在场地原是一片工业区，用于存放大量的机动车救援设施。多伦多市政府在收回这块场地后，决定在这里建一个公园，该设计公司参加了这个颇具挑战性的工作。

该项目场地比较复杂：加拿大国家铁路线经过公园的南侧地带，公园的西侧是一个救援队站点。由于缺少服务设施，设计公司不得不与多伦多市政工程部合作，将输水管和排污管从街区延伸到公园内，并运来大量干净的土方代替被污染的土层。设计师在公园与工业区之间种植了很多树木，形成一个天然隔离带。

该项目的规划包括一个初级的游乐设施、休息区、开放的草坪和很多四季常青的苗圃。场地现有的道路尽头是回车线设施，周围是可移动的护栏。通往公园的道路两旁设置着坐椅、花坛和自动饮水器。

This former industrial site was once home to an intensive auto wrecking facility. PMA became involved when the City gained possession of the land to create a neighbourhood park on this challenging piece of land.

The site was complicated by a CN Rail line running along the entire southern edge of the park and an auto wreckers on the west side. A lack of servicing required that PMA coordinate servicing of the site with the City of Toronto Works Department, extending water and sewer lines to the park from the existing street. The contaminated soil on the site was excavated and placed in large berms which were than capped with clean imported material. The berms were heavily planted with forest like material to create a natural buffer between the park and the surrounding industrial users.

The park design includes a junior play structure, nearby seating, open fields of grass for unstructured play, and perennial beds. The existing road was given a new terminus with turn-around facility beyond removable bollards for municipal use. Decorative paving provides a sense of entry to the park with its seating, flower beds, and drinking fountain beyond.

科尔曼公园

Coleman Park

撰文／图片提供：Ground Inc.　　翻译：谷晓瑞

该项目位于佛罗里达州西棕榈滩的非洲裔美国人聚居地。这里有着悠久的历史——其前身为林肯公园，于1918年成立的国家黑人联盟棒球队将这里作为主要的活动场所之一。在该公园最辉煌的时期，孩子们在这里嬉戏，社区里的很多活动都在这里举办，这里俨然成为了周围社区的活动中心。然而，随着时间的流逝，该公园和附近的社区渐渐变得杂乱无章。

该项目的整修和科尔曼公园活动中心的建设使公园再次成为社区的活动中心。在这里，经常会举行对成年人宣传教育以及孩子们的课后讲座等活动，附近

的居民每天都被吸引到这里。这片户外空间不仅供孩子们玩耍，也成为了社区的社交中心。

这片广阔的户外空地如同一张空白的帆布，设计师在上面加入了艺术元素，在继承原址传统文化的同时又不乏律动的气息。

在国家黑人联盟棒球队曾经打球的地方，设计师按照棒球场地基线的位置和场地布局，用碎石铺设小路。场地里分散地矗立着9座巨型棒球石雕，以此作为对球队的纪念。直径约1.2米的雕塑要比实际棒球大很多，刻画得十分精细，深受人们喜爱。长凳、舞台、桌子、背景幕布成为了孩子们攀爬嬉戏的欢乐天堂。为了纪念棒球手的事迹，人们提议将每年社区中推选出的社区英雄的名字题刻在棒球上。一段段的英雄事迹将被记载在这片场地上，以此将过去和现在联系起来。

该项目的设计包括整座公园的循环系统和路径、雕塑的设计及其位置安排。在该项目的设计改造过程中，社区的居民发挥了积极的作用。设计师根据社区中心夏令营活动中青少年的报名数量确定每个棒球的位置，并设置了全尺寸的纸板模型。

当地居民为该项目的设计付出了自己的力量，共同打造了这个公共场地，体现了社区的优良传统。该项目的设计理念十分独特，尽管出发点是纪念国家黑人联盟棒球队，但设计师并没有局限于怀念过去，而是对街区进行了修缮，使当地居民对此也引以为荣。设计巧妙地将其辉煌的过去融合到现代的公共活动当中，使公园成为了社区缅怀历史的中心地。

错落而设的雕塑突出了社区动态发展的特点。棒球上每增加一个名字，就会相应增加一层板以记录其事迹。该项目记录了各个时期的英雄，并将社区悠久的历史文化与社区的发展和憧憬紧密地联系在了一起。

Coleman Park is located on a site with a rich history within the African American community of West Palm Beach, Florida. The site, previously known as Lincoln Park, served as a key playfield for the National Negro Baseball League, formed in 1918. At its peak, the park was central to the community, as a place for children's recreation, community events and fairs. Over time, the park and the surrounding community fell into disarray.

The revitalization of Coleman Park and the construction of the Coleman Park Community Center will restore the site's central role within the community. Multigenerational programs, including adult education and after school programs, bring area residents to the park from day to night. The outdoor space not only supports children's play but also provides a central area for social gathering.

The designer took the charge of taking a blank canvas—an undefined open space —and transforming it with an artistic intervention that honors the heritage of the site while encouraging interaction and active play.

The work recalls a site where the African-American Baseball League played by plaiting the field with stone-dust pathways—reminiscent of the baselines and mow patterns of a baseball field. Scattered across the field are nine oversized cast stone baseball sculptures giant, embedded in the ground at different levels. The 4-ft diameter sculptures are detailed with stitching patterns, but at their exaggerated scale depart from their original reference and become props in the theater of everyday; benches, stages, tables, backdrops. They have become a beloved feature for children to climb, play and gather around. The community is encouraged to take ownership of the sculptures. Reminiscent of ball players' autographs, it is proposed that each year the neighborhood honor community heroes by inscribing each name on one of the baseballs. Layers of stories will be written on the field over time, connecting past and present.

Community members played an integral role in the design process. The location of each baseball was determined with the input of children enrolled in the Community Center's summer program, setting full size cardboard "mock-ups" in the field.

The scope of the submitting landscape architect included design of circulation and pathways through the site, design

of the sculptures and their placement. The landscape architect worked with a local landscape contractor to install the landscape components of the project, including stone-dust pathways, and foundations for the sculptures, as well as a fabricator to build the cast stone sculptures. In addition, a structural engineer was sought to determine appropriate depths for the sculpture foundations.

The design of Coleman Park is unique in expressing the heritage of a community while engaging residents in building a shared place. Although the springboard for the design was to honor the African-American Baseball League, the park does not merely memorialize the past. It restores the core of the neighborhood and allows residents to reclaim and take pride in a place. By recognizing the rich history of the site and incorporating public involvement in the design process, the park has become a central piece in the shared history of the community.

The sculptures in their varied relationship with the ground emphasize the dynamic nature of an evolving community. With each inscription added to the baseballs, a new layer of narrative is recorded. The park honors the heroes of past and present, linking the community's rich cultural heritage with growth of the neighborhood and its continued revitalization.

国营昭和纪念公园

Showa Kinen Park

撰文 / 图片提供：日本 ATLAS 设计公司

该项目位于日本东京都西部多摩地区的立川市和昭岛市，占地面积 1800 000m²。

1977 年（昭和 52 年），美国将"美军立川基地"返还给日本政府，同年也是日本昭和天皇继位满 50 周年之年。因此，日本政府决定在"美军立川基地"遗址上建造一个国营公园，以纪念天皇继位 50 周年，公园由此取名为"国营昭和纪念公园"。

1983 年（昭和 58 年），日本国营昭和纪念公园一期盛大开园，之后的改建、调整、修缮过程一直延续到今日。

在公园规划之初，由专家委员会研究讨论确定了公园的主题——"绿色环境的再生、人性的升华"。

因此，将公园的规划设计重点放在了如何营造宽阔、开敞的绿色公共空间，以及如何合理配置具有文化内涵的公园设施上，即无论是现在还是以后都要承担起教育公众在自然环境中健全身心、培养智慧的重任，并能够提供多样化的休闲娱乐空间。除此之外，该项目作为发生巨大震灾、火灾时的市民避难场地，有着不可或缺的空间要求，这一点也是公园规划设计的必要条件。

该项目自然环境优越，特别符合武藏野地区立川森林规划的设想，因此设计师将其设计目标锁定在创造舒展放松的场地空间上。

交通系统的规划充分考虑了周边地区人们的出行条件。设计师以公园的设计理念为依据，考虑游人在休闲、健身、修身养性、避难等多方面的需求，因地制宜地进行设计。原本平坦的场地通过地形的堆砌，使景观具有丰富的变化，每个广场根据地势展现出不同的姿态，形成了该项目独一无二的景观特色。植物配置方面集合了避难、修景、展示、保护、管理等因素对植物的要求，尽可能使功能与美观相互平衡。

该项目的设计重点当属园内一座古朴的日式庭园设计，它传承了日本传统的审美观，是热爱日本文化的人们首选的切磋交流场所。

园中的欢枫亭是日本传统的数寄屋式建筑，其建设严格遵照传统的工艺技法。目前，欢枫亭已作为茶室向公众开放，游人可以观赏体验正宗的日本茶道和花道文化。

以欢枫亭为核心修建的庭园采用了池泉回游式的造园形式，将山、川、林、海等自然元素巧妙地写意于园林之中。整个庭园充分体现了日本人民崇尚自然、钟爱朴实的平和心境。

Showa Kinen Park is located in Tachikawa and Akishima, Tama Area, western of Tokyo, with an area of 180 hectares.

In Showa 52nd Year (1977), America returned the "US Army Tachikawa Base" to Japan, while Showa Emperor took the throne for 50 years. Japan Government decided to build a national park named "Showa Kinen Park" on the site of "American Army Tachikawa Base" to commemorate the 50th Anniversary of Emperor Showa's reign.

In Showa 58th Year (1983), the first phase of Showa Kinen Park was grandly opened and has kept pace with times through reconstruction and adjustment till now.

In the initial plan, the Committee of Experts formulated the theme positioning of the park after research and discussion is "Regeneration of Green Environment, Sublimation of Human Nature".

The design of the park focuses on how to build a broad spacious green public space and how to configure park facilities full of cultural connotation. The park has to undertake such responsibilities as educating nationals, promoting physical and mental health, cultivating wisdom and providing diversified recreation activity space. Besides, it is also used as a public refuge place for citizens in huge earthquake and fire. Showa Kinen Park has an indispensable space condition in planning design.

With advantageous natural conditions, Showa Kinen Park accords with Tachikawa Forest Plan in Musashino region. The design objective focuses on the theme of building a comfortable and relaxing site.

The traffic planning takes full consideration of travel

conditions in the surrounding areas and tourist demands in recreation, fitness, self-cultivation and refuge, and classifies areas according to the public planning concept. Original flat site after piling up has various changes in sight and shows different postures according to the terrain, forming a unique landscape. Lawns, grass, flowers, bushes, arbors, brooks, sculptures, gardening and background music are harmonious in the premise of width, height, and depth, dramatically rendering the seasons change all the year round and the time running in a day. Virescence arrangement integrates the refuge, scenery, display, protection and management requirements to plant and achieves an ideal balance of aesthetics and functionality as much as possible.

The design points lie in the primitive Japanese-style garden design, which inherits the historical traditional aesthetics of Japan as an ideal communication place for ones loving Japanese culture.

Happy Maple Pavilion is a traditional Japanese Sukiya building strictly complying with the conventional crafting techniques. Currently, Happy Maple Pavilion is opened as a teahouse to appreciate the genuine Japanese teaism and flower culture.

Japanese Garden centered on Happy Maple Pavilion is built in a gardening style of tortuous streams, and collects mountains, rivers, forests, seas and other natural elements, giving full play to Japanese citizens' moderate mind advocating nature and simplicity.

"花树"建筑打造的自然兰花园与通透围墙 —— 麦德林植物园

Natural Orchid Garden with Flower-tree Structure and Permeable Wall — Botanical Gardens of Medellín

撰文：Jimena Martignoni　　图片提供：Carols Tobon　Felipe Mesa　Alejandro Bernal　J. Paul Restrepo　Camilo Restrepo　　翻译：刘建明

1　兰花园鸟瞰图
2　由六边形模块构成的"花朵"屋顶
3　不规则的兰花园屋顶

总平面图

"花树"建筑平面图

自 2005 年以来，哥伦比亚麦德林植物园经历了一次脱胎换骨的变迁。作为城市改造计划与植物园内部规划的一部分，相关部门做出了两个意义非凡的决策：第一，改造整个园区的实心围墙；第二，改造兰花园。

第一个决策拉开了园区改造的序幕，但它并不仅限于对园区围墙进行改造这一看得见的层面，它也是向外界展现与新千年宏伟目标密切相关的整体规划的象征。前期开展的一系列具有战略意义的项目以及其他一些处于规划阶段的项目，所有这些无不昭示着这一饱经沧桑的城区的改造工程正式启动。希望公园（Parque de los Deseos）是一处市民进行文化交流和休闲娱乐的新的公共休憩场所；北方公园（Parque Norte）是趋向于田园风情和经典情怀的公园的典范；而探索公园（Explora）则是集互动式博物馆和开放空间于一体的文化景点。正如当地居民对拆除原有实心围墙行为的评价——围墙的拆除不仅为园区营造了一

1、2 "花树"建筑
3 兰花园中茂盛的植物

个新颖且独具匠心的通透的屏障，同时也开辟了许多新的公共区域。拆除围墙的惟一目的就是拓宽植物园的人行道，从而构建出灵活的、易于感知的场所。

第二个决策大大增强了植物园规划的美感。原有兰花园是金属结构的，屋顶与传统的建筑紧密相连，为展览提供了便利。而如今的兰花园则堪称是应用了系统理论的设计典范：用六角形模块来构筑形状不规则的复合体，并在距地面20m高处勾勒出"花树"的概念图形。

麦德林植物园始建于1972年，适逢第七届国际兰花会议召开。19世纪90年代后期，植物园的原址还是一个公共洗浴场所；直到1913年，为纪念该省独立100周年，此处才被改造成一座名为"独立森林"的面向公众开放的自然公园。1985年，植物园申报麦德林文化遗产，并于1989年加入植物保护国际组织（简称BCGI）。20世纪90年代，由于长期严重疏于维护，政府甚至考虑关闭植物园，不再对外开放。但幸运的是，在植物园所有者和政府的共同努力下，2003年植物园被列入到城市改造计划之中。

Lorenzo Castro和Ana Elvira Velez受委托为植物园设计新颖的、通透性强的围墙，前者是一名长期从事城市公园项目的波哥大建筑师，后者则是一直致力于规划动感十足的麦德林城市景观的本土建筑师。虽然围墙改造早于兰花园项目，但却在兰花园项目之后竣工。这是由于围墙的改造工作涉及到公共区域的重新规划，而且由重新规划滋生的各种各样的客观因素也导致了施工时间超出预期。如今，焕然一新的围墙俨然成为植物园最吸引人的景观元素，远远地即可领略到植物园婉约葱郁的风景；连贯的高低起伏的金属丝网被统一漆成黑色，与植物园波浪形的地势相映成趣。事实上，这一景观元素也是在向世人宣告着植物园新的开始及其与这个城市的融合。

与围墙项目不同的是，兰花园项目为本地设计竞赛的标的物。中标团队由两家设计事务所组成，这两家事务所对组织形态和自然系统有相同的兴趣。一方面，由年轻景观设计师Felipe Mesa和Alejandro Bernal带领的"B计划"事务所擅长在设计中利用地形与自然景观元素；另一方面，生于不同年代的J. Paul与Camilo Restrepo擅长将一些好的想法与创新理念完美融合。在西班牙巴塞罗那取得建筑学、城市化与城市文化硕士学位的Camilo Restrepo解释说："当看到破败的兰花园的时候，我们立刻联想到建筑与自然的完美融合。我们想在这个项目中完美地呈现自然体系中有

3

代表性的形状，"他又补充道，"我们从一开始就想到了重复性的模块以及树的造型。"

兰花园新的设计方案能否脱颖而出取决于表现手法和灵活性这两个重要因素。园内的建筑不仅要能遮蔽种植兰花的特定场所，还要为公众活动提供灵活的空间，包括一些小型的服务区，此外，建筑还必须能在较短的时间内搭建起来。

建筑的基本形状为边长 4.8m 的六边形，7 个六边形模块组建成巨大的"花朵"：一个模块位于中间，另外 6 个模块像花瓣一样环绕在周围。与上述"花朵"一模一样的复合体搭建在一起，构成独立的屋顶结构，其外缘和总体轮廓虽然呈不规则状，但又有机地联系在一起，体现出受到自然启发的特点。当 10 朵"花

以复杂而扭曲的形式延伸到地面时，又演变成了"树"，这种复杂而扭曲的结构中的每一个部分又成为更小的似乎有生命的结构。这些面积约 20 ㎡ 的空间其实就是兰花的容器，是屋顶平面与地面之间的过渡空间。

从靠近地面的第一个六边形模块起，"树干"的每一个六边形模块自下而上按逆时针方向略微旋转；整体结构上覆盖有精制的木杆，木杆之间轻微错位，增强了整体设计方案所突显的动感理念。

实际展览区是以六根纤细的金属柱为骨架的开放空间，高度为 6m，确保与人的身高有合适的比例；明暗反差强烈的大型热带蕨类植物与喜阴植物构成了花园般的空间，它们交替出现，你中有我，我中有你，在屋顶结构下营造出一系列富有韵律的自然景点，为

在此间散步或休憩的游客、嬉戏玩耍的孩童们提供了足够宽敞的空间；每逢节日或社交盛事，如音乐会、食品展览交易会、地方庆典或者音乐派对，人们似乎很愿意置身于这处洋溢着生命活力的空旷空间。

这些"花树"结构仿佛是生活在同一现实空间的生灵。事实上，在提到兰花园时，人们会自然而然地谈及这些有趣的直观参照物。"就像是在蜂房里一样"是人们说的最多的一句话，这已经让设计者感到了莫大的满足："在谈到对兰花园的印象时，受访者回答频率最高的一个词就是'惊讶不已'，所有的描述也都是围绕这个词展开的"。

为了在兰花园内设置一些服务区，如卫生间和园区管理处等，设计师在兰花园不易被人觉察的地方建

造了两座小型建筑，同样采用六边形模块进行布局，覆盖有同样的木杆，建筑物也有趣地按一定形状排列——封闭的六边形和开放式六边形交替出现，构建出景观化的娱乐休闲景点。

兰花园在竣工一年半之后才对外开放，这里已经成为这座城市尤其是其北部地区的地标性建筑，而在几年前，这座城市的上层名流根本不屑于参观这样的景点。今天，植物园向麦德林全体市民开放，园中有人们倾力保护的自然生命，更值得一提的是植物园已经成为招揽外来游客的优质景点。不久之前，植物园又新增加了一个入口、咖啡厅、植物标本馆、蝴蝶亭和杜鹃花庭院，上述项目都处于最终完善阶段；新通道的修建以及湖泊与花卉区的改造进一步完善了园区的整体规划。每年 8 月，麦德林最具吸引力的重大庆典之一——花卉节开幕时，植物园被装扮成一条展示色彩和形状、充满芬芳的自然长廊。毋庸置疑，植物园围墙改造和新兰花园的建造堪称是英明的决策；惟愿园区中的在建项目能够和上述项目一样成功，切勿成为孑然独立的外来之物。

The Botanical Gardens of Medellín in Colombia, have been experiencing a profound change since 2005 when, as part of the city renovation programs and also a holistic internal plan, two decidedly significant decisions were made: one is the replacement of the solid wall that closed the entire site's perimeter, and the other is the renovation of the "orquideario" or orchid garden.

The objective of the first decision was the opening of the site, not just visually but also as a sign of social openness and integration that was deeply related to the ambitious goals this city had set for the new millennium. Many strategic projects had been implemented in the area and others were in planning phases, all of which had started out as a process of rehabilitation of a historically quite degraded area. Parque de los Deseos, as a new gathering public space for cultural and recreational activities; Parque Norte, as an example of a more bucolic and classic park and Explora, as an integrating project where different interactive museums and open spaces make up one single cultural spot. "The fall of the wall", as the locals refer to the removal of the former hard wall, meant not only the incorporation of a new, specially-designed, permeable screen to the site but the addition of new public open areas for this part of the city. Thought of the sole purpose of enlarging the adjacent sidewalks of the Botanical Gardens, the fence's original tracing was considerably modified and this generated a flexible, easy to perceive, new sense of place.

With the second decision, the Botanical Gardens made a significant step forward regarding aesthetics. The previous orchid garden was a large metallic structure whose roof established a clear connection with those typical quite impersonal Botanical constructions that only seek for the provision of smooth exhibiting areas. What is the "Orquideorama" now (the so-called "Orqui") arises as a paradigm of the application of systems theories to design: a fractal composition whose basic hexagonal module repeats itself as many times as necessary to create the conceptual image of a "flower-tree", opening up 66 ft above the floor.

The Botanical Gardens of Medellín were founded in 1972 to coincide with the convening of the 7th International Congress of Orchids. Originally, in the late 1890s, the site had been a public bathing place until 1913 when it was transformed into a public nature park called Bosque de la Independencia or Independence Woods, in honor of the first centenary of the province's independence. Later, in 1985 the Gardens were declared Cultural Patrimony of Medellín and in 1989 joined BGCI or Botanic Gardens Conservation International. During the 1990s the site suffered a considerable general lack of maintenance and closing it to the public was even considered an option. Fortunately, in a laudable effort

of the Gardens' owners and the City Government, in 2003 the site turned into the object of an important renovation that commenced with the two aforementioned decisions.

The design of a new permeable fence for the entire site of the Botanical Gardens was directly commissioned to Lorenzo Castro, an architect from Bogotá who had been working on urban parks projects and Ana Elvira Velez, a local architect who has been into Medellín's current very vibrant urban scene since its beginning. Although this commission was previous to the one of the Orchid Garden, it was completely finished later; the fact that this project involved the redesign of public areas, and consequently a larger and more diverse number of actors, led to a longer total execution time. Today, the new fence appears as an inviting element through which the handsome greenery of the site can be fully appreciated from the outside; made with consecutive wire-meshed panels painted in black and whose tops seem to follow the undulating surfaces of the gardens, this element becomes the first real statement of the site's opening and integration with the city.

On a different note, the new Orchid Garden was the object of a local design competition. The winning team was formed by two offices who shared the same interest in organic shapes, systems of nature and phenomena. On the one hand, Plan B, an office led by Felipe Mesa and Alejandro Bernal, two young architects who had been experimenting with topography and nature in their design projects; on the other hand, J. Paul and Camilo Restrepo, two other architects who, embodying two different generations, came to a fine integration of ideas and innovative concepts.

Camilo Restrepo, who completed a Master degree of Architecture, Urbanism and Urban Culture in Barcelona, Spain, explains: "When we saw the old orchid garden we were very attracted to the sense of integration between architecture and nature. We wanted to apply the kinds of shapes that were well represented in natural systems." Then he adds: "We started to think about a repetitive module and also about the shape of a tree as a generating one."

The conditions for the new design resided in two key concepts: representation and flexibility. The structure had to be able to shelter not only specific places for orchids but also to offer a flexible area for social events, including some small service areas. In addition, it had to be quick and easy to build.

The basic figure is a hexagon whose side length is 4.80 meters and the "flower" is a composition made of seven modules, a central one and six more which make up what alludes to petals. This composition is then repeated ten times in order to generate one single roof structure whose external edge and general silhouette is therefore irregular, organic, nature-simulating. The ten "flowers" actually become "trees"

when they reach the floor through complex, twisted bodies, each of which becomes some smaller structures or "living yards". These spaces, of approximately 20m², are the actual orchids' containers and, structurally, define the transitional space between the plane of the roof and that of the floor.

Each hexagonal module in the "trunk" is rotated slightly counterclockwise from the one below and crowns the orchids and green composition; the entire structure is covered with fine wooden rods which, slightly shifted with respect to each other, strengthen the idea of motion that seems to have engendered this whole design.

The actual exhibiting space, framed with six metallic svelte columns, remains open and has a height of 6 meters which guarantees a correct relation to human scale. Large tropical ferns and shade plants – which go from unbelievably bright to relaxing dark green hues – compose these garden-like spaces; they alternate, hide each other, embrace each other, creating a series of natural spots that seem to rotate underneath the repetitive, almost cyclical, roof structure. However, they also create a rhythm and mark the space, leaving a large flexible area where people walk or sit and kids run or play around; during festive days and social events (concerts, food fairs, local celebrations and even electronic dance music parties) the crowd seems to adapt to the empty areas shaped between those vertical life-exuding elements.

These "flower-tree" structures appear as living creatures that share one common fractal space. In fact, people come up with interesting visual references when they talk about it. "It's like being in a honeycomb" is one of the most typical comments and, actually, one that makes the designers feel really satisfied. According to them "The possibility of having people expressing their own perception of the Orquideorama is amazing. All their images relate to a language that is common to all of them."

In order to provide some service areas, such as sanitary rooms, administration and private offices, the Orquideorama offers two subtle constructions that want to go unnoticed. Laid out with the same generating hexagonal module and covered with the same wooden rods, the arrangement of these buildings also evokes an interesting play of forms; closed hexagons, which respond to the functional program, alternate with open ones, which create landscaped relaxing spots.

More than a year and a half after its completion and opening to the public, the Orquideorama has become a landmark of the city and, especially, of the north part of it, where no more than a couple of years ago people from upper classes would not have even thought about visiting. Today, the Botanical Gardens open to Medellín's citizens with no distinctions, share the natural life they protect, and, additionally, become a point of interest and attraction for tourists. The place has also recently incorporated a new entry building and café, a herbarium, a butterfly-pavilion and a rhododendron yard or "patio de azaleas", all of which are in final phases; new paths, lake renovation and revision of the flora collection complete the general plan for the site. In August, the month in which "la Feria de las Flores" or Flowers Fair takes place, one of Medellín's major and certainly attractive celebrations, the Botanical Gardens turn into a natural gallery full of color, shapes and perfumes.

The redesign of the site's fence as well as a new orchid garden has proven to be undoubtedly good decisions; hopefully, the proliferation of new buildings which are now being implemented within the site will prove to be the same and won't stand only as autistic objects inside the green.

1 植物园的服务区
2 改造后的通透围墙

岬角公园

Ballast Point Park

撰文 / 图片提供：McGregor Coxall　　翻译：王玲

该项目是一座景色优美的滨水公园，它由代表新南威尔士州政府的悉尼港海岸管理局发起、McGregor Coxall 景观设计事务所设计而成。该公园于 2009 年 7 月对外开放，它是社区活动的成果，使原本用于开发住宅的地块摇身一变成为惠及于民的公共公园。McGregor Coxall 景观设计事务所领导着一个多学科的设计团队，将加德士公司在悉尼港上的储油炼油厂转变成一处占地 28 000m² 的公共公园。

该设计采用先进的可持续性发展理念，尽量减少碳足迹，恢复地块的生态环

境。场地的历史层次与前瞻的新技术完美融合，形成了一个重要的区域性城市公园。随处可见的雨水生物过滤系统、再生材料和用于地块能源供给的风力涡轮机无不展现出该设计的环保理念。

该项目的设计挑战了设计师对于材料及其使用的传统认知。新建的高低起伏的挡土墙高高地矗立在砂岩峭壁之上，这些墙体没有采用别处的珍贵砂岩，而是利用另一块场地上原有的碎石修筑而成。曾经无用的压舱碎石不仅变成了漂亮的景观元素，还成为整个石笼挡土墙的构筑主体。墙体顶端是带有精致护栏的混凝土板，这一切都使碎石墙体成为面对内港入口处的一道靓丽风景线。

八个垂直轴风力涡轮机以及镌刻在回收的储油罐饰板上的莱斯·穆雷的诗歌重新诠释了场地上原有的

最大储油罐。风力涡轮机象征着未来，它标志着人类由过去的化石燃料转向可持续性的再生能源形式的进步。主要的设计理念是：

1. 展现出地块从过去的工业腹地到如今的公共公园的蜕变，使曾经粗犷的工业场地流露出些许的精致与典雅。

2. 展现出人们对景观文化认识的变化，这是一种从资源利用到资源尊重的态度的转变——探索性地利用场地上原有的碎石建造了一座未来公园；场地回收材料的再利用和场地能源需求的自我开发进一步体现了这一理念。

3. 尊重和展现场地的历史层次，这一点在与场地历史层次和谐相融的设计中被恰到好处地体现出来。

4. 可持续性设计理念被运用于场地雨水和植被的管理中。场地上所有的雨水都被引入种植区，雨水在这里经过净化和过滤之后流入港湾。

最初，总体规划强调拆除与保留之间的精准平衡，而该设计则为它注入了新鲜的活力。公园以创新而有效的方式使场地的历史层次与人类的干预活动彼此交融、相得益彰。最终，这里将绿意盎然、身披"戎装"。

该项目涵盖了所有景观原则的基本概念，也是长期不断的社区咨询与场地遗产研究共同作用的成果。其设计策略对建筑技术和材料进行了许多创新的尝试，包括附有碎石石笼、重新加固的挡土墙，再生材料替代传统材料的环保混凝土，为建筑和景观小品而回收利用的木材、土壤、护根层和砾石。场地植被也多采用本地植被，不仅促进了当地的基因库，同时对重建当地的动物群落也具有一定的意义。当这里植被葱茏之时，又会与帕拉玛塔河入口处的 Balls Head 岬角形成一对遥相呼应的姊妹岛。

Ballast Point Park is a stunning new harbourside destination, delivered by Sydney Harbour Foreshore Authority on behalf of the NSW Government and designed by McGregor Coxall. The park is a result of community action that stopped development of the site for residential development and returned the land to people of Sydney as a parkland. Ballast Point Park was opened to community acclaim in July 2009. This project involved McGregor Coxall leading a multidiscipline team in developing the design for this 2.8 ha new public park on the former site of a Caltex oil storage and grease manufacturing plant on Sydney Harbour.

The design uses world leading sustainability principles to minimize the project's carbon footprint and ecologically rehabilitate the site. The design reconciles the layers of history with forward looking new technologies to create a regionally significant urban park. The environmental approach is further underpinned by site-wide stormwater biofiltration, recycled materials, and wind turbines designed for on-site energy production.

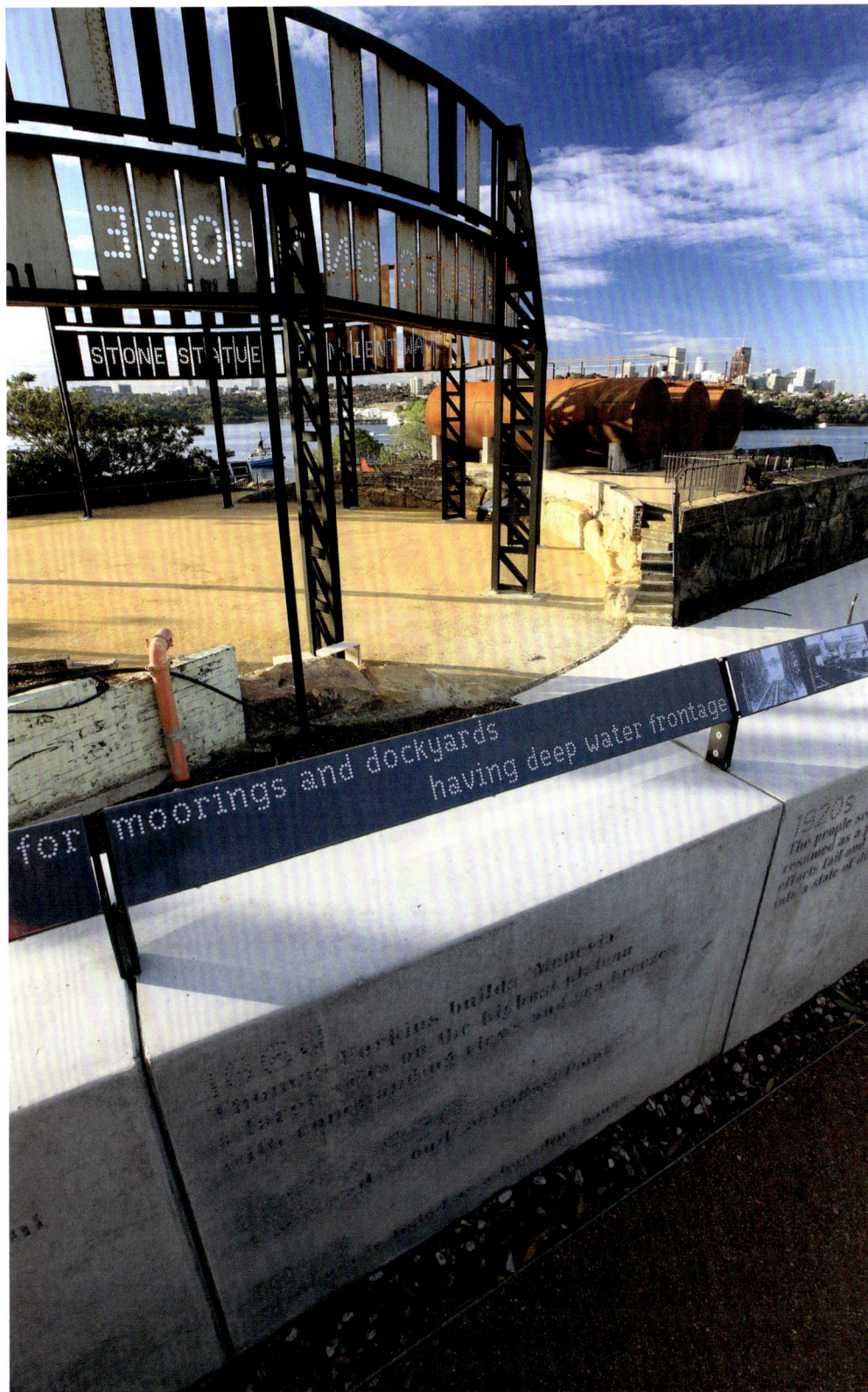

The design challenges our perception of materials and their use. Dominant new terrace walls sit atop the sandstone cliffs but these walls are not made of precious sandstone excavated from another site, rather from the rubble of our past. What once was called rubbish is now called beautiful. It is the new ballast. But it is more than this at play: It is the total composition of these recycled rubble filled cages, off set with concrete coping panels topped with fine grain railing, that allow these walls to sit confidently at the portal to the inner harbour.

8 vertical axis wind turbines and an extract from a Les Murray poem, carved into recycled tank panels, forms a sculptural re-interpretation of the site's former largest storage tank. The wind turbines symbolise the future, a step away from our fossil fuelled past towards more sustainable renewable energy forms. The underpinning philosophies for the site are:

1. To communicate the change that has occurred to this unique area of Sydney, i.e. from an industrial heartland to its present more gentrified evolution. The design overlays a high end fine grain detailing over a more robust constructed base.

2. To communicate the change in our cultural perception of our landscape from a resource to be used to an asset to be respected. The design explores the re-use of the rubble of the past in re-building a park of the future. This concept is further developed by exploring the use of recycled material across the site and the generation of its own

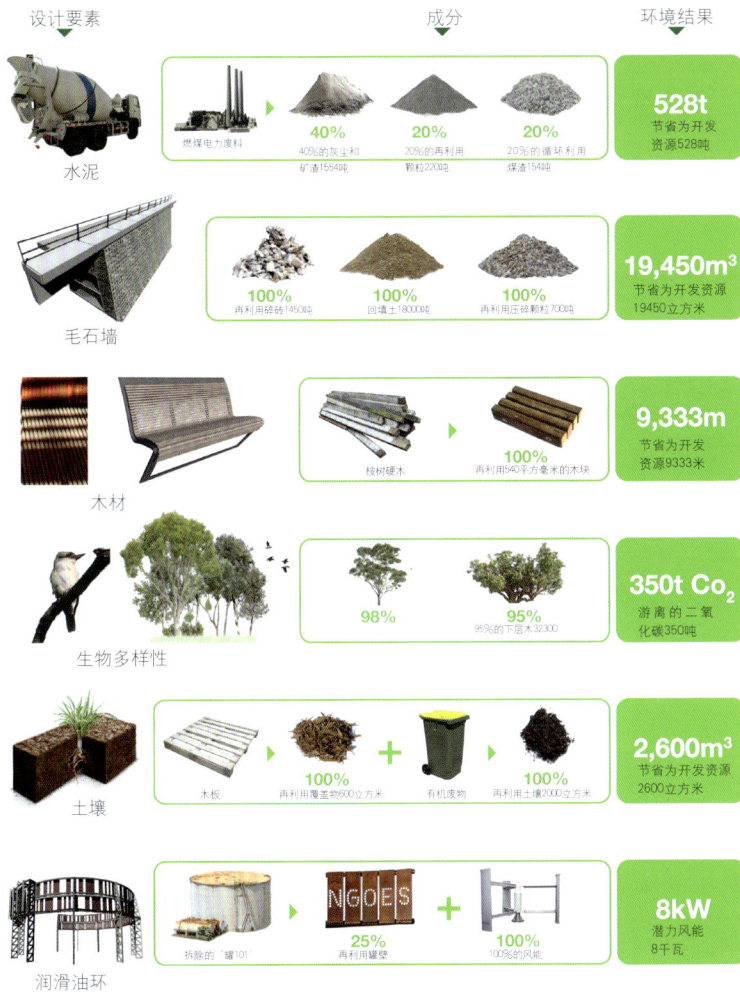

设计要素 | 成分 | 环境结果

水泥
燃煤电力废料
40% 40%的灰尘和矿渣15540吨
20% 20%的再循环颗粒220吨
20% 20%的循环利用煤渣1540吨
528t 节省为开发资源528吨

毛石墙
100% 再利用碎砖14500吨
100% 回填土18000吨
100% 再利用压碎碎粒700吨
19,450m³ 节省为开发资源19450立方米

木材
桉树硬木
100% 再利用540平方毫米的木块
9,333m 节省为开发资源9333米

生物多样性
98%
95% 95%的下层水32300
350t CO₂ 游离的二氧化碳350吨

土壤
木板
100% 再利用覆盖物600立方米
有机废物
100% 再利用土壤2000立方米
2,600m³ 节省为开发资源2600立方米

润滑油环
拆除的"罐101"
25% 再利用罐壁
100% 100%的风能
8kW 潜力风能8千瓦

energy requirements.

3. To respect and communicate the site's historic layers. This is achieved by the site's historic layers being finely interwoven into the design.

4. To employ sustainable design principles in managing on site stormwater and planting. All site stormwater is directed to planting areas where it is cleaned and filtered prior to entering the harbour.

This design brings to life the principles established in the original master plan where a fine balance between what is removed and what is retained is proposed. The end product is a park that proudly communicates all the site's past layers and human interventions in both, an innovative and informative manner. However, the true story line here is that ultimately the planting strategy will result in this headland being re-clothed in green.

This project embraces the underlying concepts of all the landscape principles. It is a result of on-going community consultation coupled with heritage research. Its design strategy explores many innovative uses of construction techniques and materials, these range from re-enforced earth walls clad with recycled rubble in baskets, green star rated concrete using recycled materials in lieu of traditional components, recycled timbers for the buildings and park furniture as well as recycled soils, mulches and gravels. The planting for the site is provenance stock drawn from local plant communities that promotes the local gene pool as well as assist in the re-establishment of the local fauna. The park provides a reference to the past and an eye to the future, when the plants are fully mature it will sit comfortably opposite Balls Head, its sister headland at the entrance to the Parramatta River.

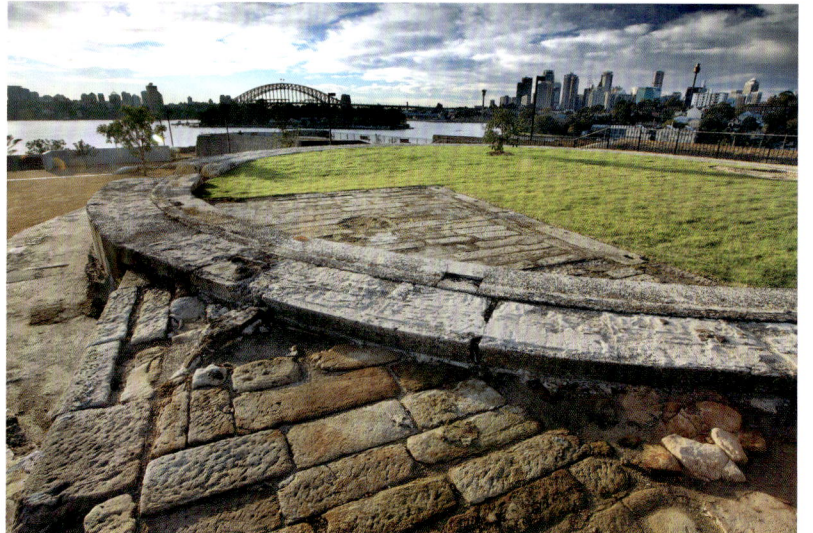

Roofo

屋顶花园

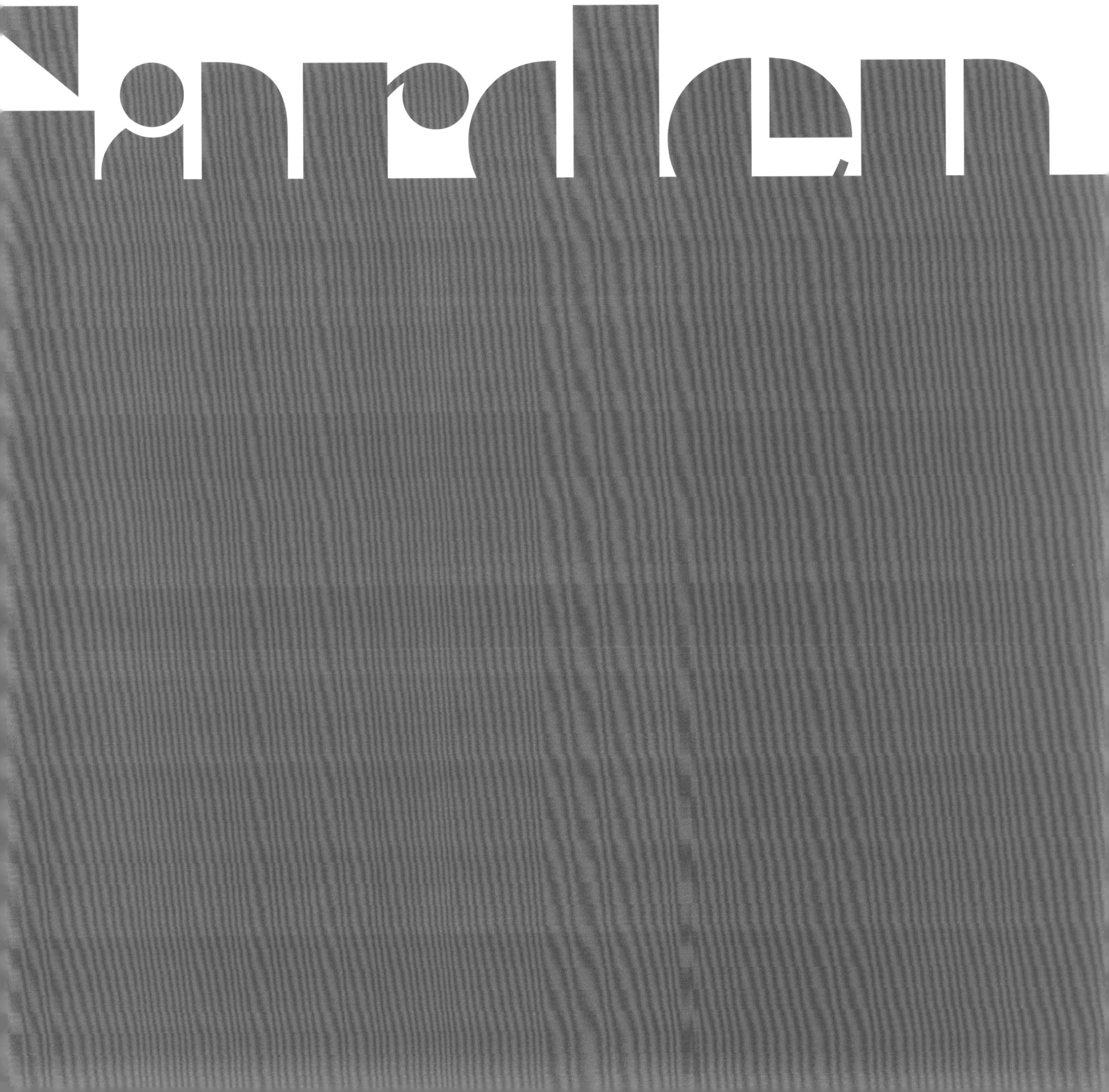

城市广场购物中心空中花园

Sky Park of City Square Mall

撰文 / 图片提供：ONG & ONG Ltd.　　翻译：张晶

设计师设想将城市广场购物中心（CSM）设计成城市边缘的"社区中心"，满足人们聚会、学习、娱乐和购物的需要。其设计灵感来源于充满活力的周边环境，与邻近的城市公园和地铁站相互融合，形成强大的枢纽中心。

购物中心发展的重要主题之一是"生态友好、与自然和谐共存"。为了实现这一主题，设计师将购物中心前方占地面积为 4500m² 的区域改造成现代公园，并将购物中心的第六层建成一座开阔的空中花园，与购物中心相得益彰，有益于提高人们的环保意识。

空中花园占据了购物中心第六层的大部分空间，人们凭窗眺望，能够观赏到周边赏心悦目的优美景色。这座空中花园采用有机的结构形式，类似于一座自然公园。简洁美贯穿于项目的整体设计之中，并预留出足够的空间供孩子们奔跑、嬉戏。花园内枝繁叶茂，人们从喧嚣的商业活动中抽身而出，置身其中享受的不只是一缕缕纯净的植物气息，更是一份宁静悠闲的心情。设计师环绕花园设置了一排长长的矮墙作为休息区，并用繁茂的植物来标示出花园的界线，以使现有空间视觉上最大化，人们可以在这里聚会、互动、享受大自然。

为了响应新加坡"花园城市"的运动，设计师克服各种难题，将场地中原有的两棵紫檀树保存下来，场地中原有的其他树木则根据造景的需要，被移植到其他位置。另外，设计师在现代公园内种植了多种植物，使购物中心俨然成为一片城市绿洲。落叶和修剪的树枝等垃圾也被收集起来，用做公园土地的肥料。

活体曲径

蝴蝶花园

绿篱

喷泉公园

态曲径

台阶

City Square Mall (CSM) was envisioned as a city fringe "Community Hub" that would provide opportunities for the community to meet, learn, play and shop. The mall's design concept draws inspiration from the vibrant surroundings, and leverages on its integration with an adjacent Urban Park and MRT station to create a strong community node.

Eco-friendliness and co-existence with nature were also key themes in the development of this mall. To this end, an expansive sky park was created on the mall's sixth level, while the 0.45 ha area fronting the building was integrated and turned into an adjacent urban park, which promotes environmental awareness.

Taking up a substantial amount of floor space, the sky park overlooks the neighbourhood, providing breathtaking views of the surrounding areas and beyond. The sky park's takes on an organic form, very like the spontaneity we often observe in nature's designs. Simplistic beauty governs the overall architecture of the landscape, and there is ample space for young children to run about and play in. Those who seek solace from the commercial bustle within the shopping mall can also relax as they breathe in the sky park's abundant greenery. Space is maximised by cleverly incorporating long seating areas into the low walls that circle the park's perimeter and neatly demarcate the lush flora that borders the area.

In line with the "City in a Garden" movement, great care was taken to avoid cutting down two existing Pterocarpus indicus. These trees were preserved and incorporated into the proposed urban park while other existing trees were transplanted to new locations on site. In addition, more trees were planted at the urban park to make it a true oasis in the neighborhood. Horticultural waste like leaves and twigs were also collected and turned into compost for use in the park itself.

高空中的休闲一隅 —— 地标

Leisure During the Sky — The Icon

撰文 / 图片提供：王及王私人有限公司　　翻译：刘建明

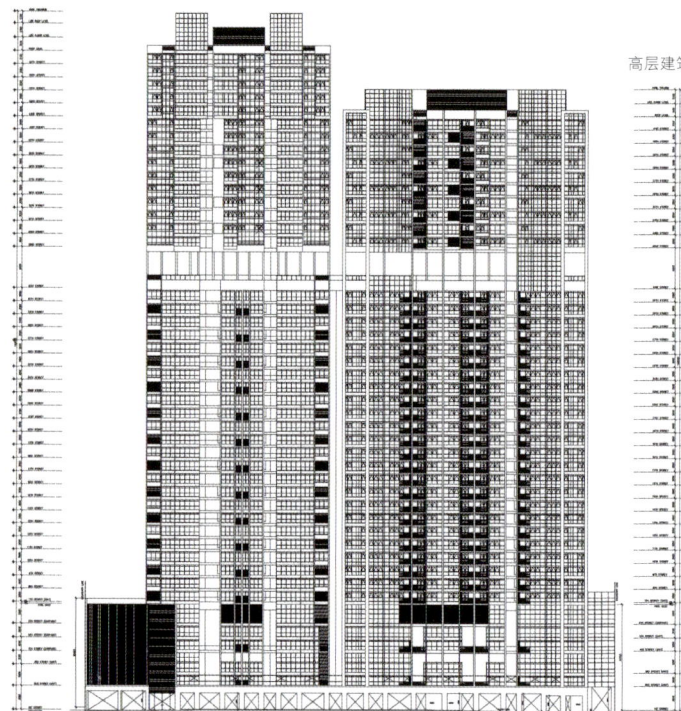

高层建筑 1

高层建筑 2

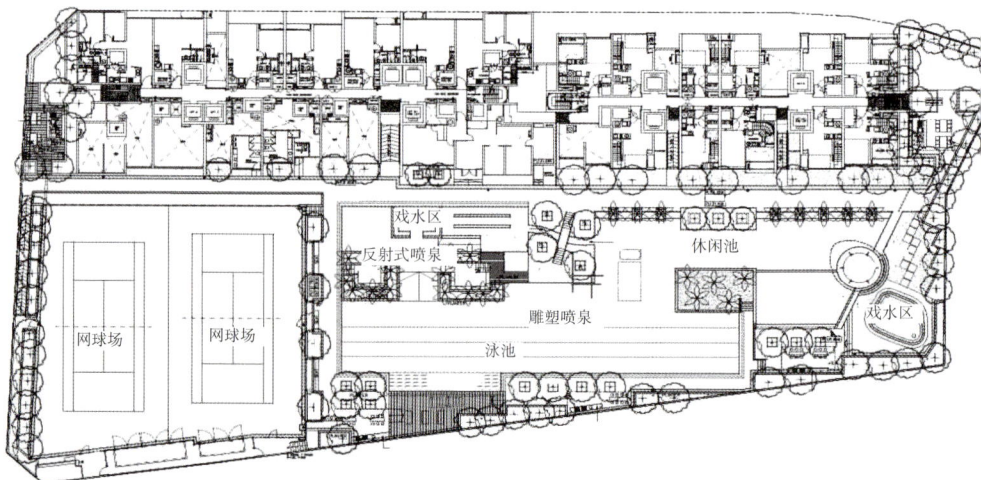

戏水区
反射式喷泉
休闲池
雕塑喷泉
戏水区
网球场
网球场
泳池

该项目是一个混合用途开发项目，与新加坡城市核心规划区内的丹戎巴葛务边大街相临，包括两座分别为41层和46层的高层建筑，其中较高的高层建筑的一楼为商场、多层停车场和其他公共设施。

设计理念

该项目专为敢于标新立异并渴望在闹市中寻得一隅栖居之所的智者而设计。在这两座高层建筑上都能够俯瞰到城市地平线和大海，而公寓南北朝向的设计又巧妙地避开了西晒。两座高层建筑之间利用空中通道连接，既节省了空间又展现了高科技的魅力。

简洁而圆滑的线条勾勒出现代风格，搭配以映现出变幻的天空画面的落地玻璃窗，为商务区平添了一丝生趣；现代建筑语言体现了都市风貌；在功能性空间与无障碍景致的完美搭配下，两层楼高的天台更显出了城市建筑群的绿色生活理念。

在项目设计阶段，空中绿化是一个非常关键的课题。该项目倾向于利用巧妙的建筑设计、创意天台、屋顶花园上的公共凉亭来突出整体的建筑设计。建在第31层的天台提供了进行娱乐活动所需要的设施。建筑物正面圆滑的线条与公寓的现代生活理念相得益彰。

一楼购物商场内部涂饰的处理采用乡村风格。墙体配以素面砖，地面采用天然石头铺设，以营造乡村气息，勾起久居闹市中人们的乡村情怀。此外，大量的独创性设计进一步增强了人们的乡村体验。即便是垃圾箱也做了精心设计，以符合乡村生活环境的特点。

公共休闲和生活便利设施

第7层上配有完善设备的洗衣房增添了社区的生活氛围：

悬空阁楼给居民提供了休闲放松的宁静空间；

残疾人专用通道通过自动旋转玻璃门和两个带摩擦带的斜坡与主大厅相连；

大厅休息区非常宽敞，可为不同用户提供足够的活动空间；

业主信箱位于坐轮椅者也可轻松触碰到的位置。

第7层的娱乐空间

该项目的第7层为娱乐区，娱乐区下面有5层的多层停车场。娱乐区为业主提供相互交流、观赏周边风景的空间。水池区占地面积约1015平方米，配置了泳道长为50米的泳池、休闲池、极可意浴缸（周边可喷水按摩的小浴池）、休息平台、雕塑喷泉、反射式喷泉和一座水上迷宫，水池区旁边设计巧妙的植物箱为角落提供了宁静而惬意的氛围；户外设施包括网球场、烤肉区、娱乐角、戏水区、日光浴平台和亭子——所有的设施都为业主间的沟通提供了便利，这也是空中绿化的一个重要组成部分。

第 31 层的天台

第 31 层的天台堪称该项目的王冠宝石。精心设计的天台为生活在"混凝土丛林"中的人们提供了一方难得的憩息之处。这里配备了极可意浴缸、热浴池、镜面水池、茂盛的绿色植物，可以说天台就是业主的天堂。木质甲板与周围的种植箱营造出柔和的环境，也增加了"混凝土丛林"的生物多样性。

第一幢高层建筑的 31 层主要是俱乐部和室内娱乐设施，如健身房、剧院、休闲室、阅读区和游戏室。

第二幢高层建筑的天台在设计和个性上并没有做任何限制。悬挂阁楼为居民提供了另外一种宁静的空间，人们可以在郁郁葱葱的绿色景观和婉约的水景中休闲放松。镜面水池上的坚固石板充当了各个不同区域之间的连接通道；天台也配备了芳香区和蒸汽室。

用玻璃屏隔开的极可意浴缸和热浴池使人们既能体验到空间的开放性，也能感受到空间的独立性。极可意浴缸和热浴池的位置靠近高层建筑的边缘，业主可以体验到绿色景观之中的高空生活。划分天台边界的玻璃面板既可以起到安全屏障的作用，同时又不妨碍业主沉浸在水池和美景之中。

灯光

停车场、住宅区走廊、楼梯和电梯中的灯光设施在白天只使用 50% 的能量。

热量

所有正面的玻璃窗都做了上釉处理，可反射午后的阳光，为建筑内部提供一个清凉的环境；

充沛的自然光透过玻璃幕墙倾泻而入；

景观平台上的水体和绿化设计旨在为地下停车场的屋顶提供间接制冷，并可降低周围的气温。

水景

儿童游乐区（例如戏水区）使用的水可供灌溉。

环保

空中绿化是该项目设计的关键部分，旨在营造一个独一无二、宜人的环境。天台提供了建筑的可持续性所能带来的健康、经济与环境效益。天台的设计既满足了业主的需求，同时也可使年轻一代懂得珍惜环境。

The Icon is a mixed-use development at Gopeng Street in Tanjong Pagar within the Downtown Core Planning Area in Singapore. The development comprises a block of 41-storey and a block of 46-storey with a commercial 1st storey with multi-storey parking and other communal facilities.

Design Concept

The Icon is designed for the sophisticated individual who dares to be different and desires a pad in the city. The Icon two linear blocks consist of 41 and 46 stories amid panoramic views of the city skyline and the sea. At the same time, it is cleverly designed to face the North and South direction in order to avoid the afternoon sun. A sky linkway is constructed to connect the two towers in an unconventionally space-saving yet sophisticated way.

A modern look with clean, sleek lines matched with full-height glass windows, reflecting the ever-changing skyscape. Icon lends an interesting addition to the business district. The architectural language echoes modern refrains reflecting the urban spirit. Built with functional spaces and unobstructed view to match, the double-storey height terrace provided on each tower emphasizes the idea of "green" living within city dwellings.

Skyrise greenery was an important component during the design phase of this development. Icon focused on clever building design and creative sky terraces and communal open pavilions in roof gardens to enhance the overall building design. The terraces found on the 31st storey will provide recreational and clubhouse facilities and the sleek lines created on the building façade complement the contemporary living concept of the apartments.

The finishes of the interior of the mall reminds one of a village. Its walls are fitted with fair-faced bricks and the floors are laid with natural stones to create and capture that nostalgic feel of village-living that is severely lacking in urban cities. In addition, there are plenty of little ingenious touches that further enhance this experience. Even the trash bins have been carefully selected to blend in with the rustic surroundings. Icon is designed to be the ultimate in urban lifestyle living. It is a home like no other - unconventional in theme yet holistic in its approach of providing unparalleled comfort.

Communal and Leisure Facilities

There is a full range of amenities and conveniences within and around the Development. These include:

Laundry room complete with ironing facilities situated at the 7th storey promotes a sense of community living.

Hanging pavilions are designed to give residents quiet spaces for relaxation.

The disabled lot is located at a strategic point that is linked to the main lift lobby via auto swing glass doors and 2 ramps with friction strips.

Lobby areas are spacious to provide sufficient space to for different users. Residential letter boxes are located at convenient levels for wheelchair bound users.

Recreational Space at 7th Storey

The Icon comprises of a recreational area on the 7th storey. This recreational space sits on 5 levels of multi-storey carpark. The recreational area provides spaces for residents to commune whilst enjoy the breath-taking views of its surroundings. The main swimming pool of approximately 1015sqm comes equipped with a 50m-lap pool, leisure pool, Jacuzzi, lounging shelves, sculptural jets, reflexology water jets and a water labyrinth. Clever use of planter boxes surround the pools providing quiet and relaxing corners. Outdoor facilities consist of tennis courts, BBQ pits, fun corner, splash zone, sun decks and pavilion — all created to facilities community interaction that is a key aspect when creating Skyrise greenery within the Icon.

Sky Terrace at 31st Storey

The crown jewel of the Icon is its 31st storey sky terrace. This sky terrace was carefully designed to provide a sanctuary within the concrete jungle. Consisting of Jacuzzi's, hot tubs and reflecting pools surrounded by lush greenery, the sky terrace is indeed a haven for residents. Timber decking with clever use of surrounding planter boxes softens and introduces biodiversity to this concrete environment.

The 31st storey located at Tower 1 consists mainly of the Clubhouse with indoor recreational facilities. This includes gymnasium, theatre, lounge and reading area as well as a games room.

The sky terrace at Tower 2 has basically no limitations in design and individuality. Hanging Pavilions are cleverly designed to provide residents with quiet spaces at a different level for relaxation amidst the lush landscape and beautiful water features. Concrete slabs lay over reflective pools to provide pathways connecting different areas. The sky terrace is also equipped with aroma areas and steam rooms.

Jacuzzis and hot tubs are separated with glass screens to allow residents to experience the openness of the space yet provide a separation between spaces. The Jacuzzis and hot tubs are also built close to the edge to give users the experience of high-rise living amidst lush landscape. An overflow weir (overflow tension edge) allows for overflowing water to be piped out from the space when hot tubs are utilized. Glass panels line the terrace serving as a safety barrier and at the same time not restricting residents from sinking into the cool waters and greenery of the haven.

Light

Lightings at the carpark, along residential corridors, staircases and lifts operate at 50% energy in the daytime.

Heat

Clear glazing on all window facades deflects heat from the afternoon sun and provides cool ambient in the surroundings.

There is a spectacular glass curtain wall that allows an abundance of natural light to stream in.

Water bodies and greenery on the landscape deck is designed to provide indirect cooling to the roof structure of the basement carpark and reduce ambient air temperature.

Water

Water used for children's play areas such as the Splash Zone is irrigated.

Conservation

Sky rise greenery is an important design element in this development. It will create a unique, welcoming environment. The sky terraces provide health, economic and environmental benefits of building sustainability. These sky terraces are designed specifically to meet the needs and lifestyles of the people who occupy them and will also serve as a learning platform for the young to appreciate our environment.

SEW屋顶花园

SEW Roof Garden

撰文：Jessica Brown　图片提供：Maria Auböck　翻译：高明

该项目是在 SEW 汽车制造公司原来的工厂内部新建一个椭圆形的培训基地，这个训练中心是厂区中扩建项目的最后一项，用以对员工进行培训。在池塘周围种植草坪、观赏花和绣球花，使新旧建筑实现了和谐统一。玻璃通道横跨池塘，把新建筑、旧建筑与新的停车场联结起来，在停车场上种植了大量的竹子和樱桃树。人们也可以在自助餐厅的阳台上观赏池塘的美景。

虽然池塘增添了人们的生活情趣，也增强了各建筑之间的联系，但工程的亮点却是美丽的屋顶花园。员工们可以在不同楼层的屋顶花园里度过休息时间，享受花草所带来的美好味觉与视觉体验，欣赏周围的美景。建筑的一层有野餐场所，还有座石砌的圆形露天剧场。建筑顶层种植着一片片蓝紫色的植物，大片的黄花和绿草点缀其中，与旁边铺设着白色大理石碎石的地面相映成趣。灌溉系统与屋顶植被和地面植被结合起来。种植床采用不规则形状的灵感来源于胡安·米罗和亨利·马蒂斯这样的艺术家。为建筑底层提供光线的圆形采光井一直延伸到屋顶花园，使屋顶看起来像一艘远洋班轮的甲板。路面都是由白色大理石碎石或木材铺成，花池边缘是由不锈钢制成。这里种植的花草种类多样，不是只有在夏天才枝繁叶茂的植物，还有很多四季常青的草本植物，确保一年四季充满生机。这些植物的位置分布受美学和屋顶风向的影响——有些地方较为隐蔽，为那些比较敏感的植物提供了更好的生长空间。鼠尾草和百里香等草本植物为花园增添了芬芳的气味；夜晚，由建筑师 Walter Holper 和 Martin Kohlbauer 设计的灯将花园点亮。这里不仅是人们感官进行享受的场所，更是人们在工作或培训之余放松心情的美好场所。

1　白色沙地上的蓝紫色植物 1
2　在自助餐厅阳台上观赏美景
3　池塘周围的草坪和花卉

总平面图

The landscape design for SEW, a motor manufacturer, integrates a new elliptically shaped training facility into the existing fabric of industrial buildings. The training centre is the last of a series of expansions on the site, and is intended as a gathering site for the continuing education of employees. The integration of new and old was achieved by using a pond surrounded by lawns and long border plantings of ornamental grasses and hydrangea. Glassed-in walkways cross the pond, connecting the new building to existing buildings and to a new parking area. The parking area is planted with cherry trees and clusters of bamboo. The pond can also be enjoyed from a terrace by the cafeteria.

While the pond adds interest and connectivity to the ground plane, the highlight of this project is the beautiful roof gardens! During their breaks, trainees can spend time on several different levels of roof gardens, enjoying the smell and colour of the plantings and a view of the surrounding area. On the first level there is picnic space, as well as a stone amphitheatre. On the top level, islands of purple and blue plantings with strong yellow accents and green

grasses are set against a ground of white marble gravel. An irrigation system is integrated with roof plantings, as well as the plantings on the ground. The free form shapes of the planting beds are inspired by artists like Juan Miro or the late Matisse; like them we play with figure and ground. The round light shafts for the building below rise into the garden, giving the space the feeling of an ocean liner deck. All of the paths are made of the white marble gravel or wood, and the boarders of the planting areas are made of stainless steel. The plant selection creates more than the lovely summer colour scheme. There are several grasses and herbs which keep their leaves in winter, ensuring that the space is lively all year. The locations of these plants are influenced by aesthetics, and also by wind patterns on the roof. Some areas are more sheltered, and provide good habitat for more sensitive plants. Herbs, such as sage and thyme, add a lovely scent to the garden. At night, the gardens are finely lit with a lighting design by Walter Holper and architect Martin Kohlbauer. Overall, the space is a delight for the senses, and a lovely spot for relaxing between training sessions.

都市屋顶与自然花园的交融 —— 百老汇684号

The Integration of Urban Roof and Natural Garden — 684 Broadway

撰文：Balmori Associates　　图片提供：Mark Dye

探索建筑物与自然环境如何和谐共生是该项目的设计目标。设计师通过空间重构和创造新的连接，营造出更流畅自然的过渡空间。然而，这不但没有模糊景观与建筑的界限，反而彰显出二者的不同之处。景观与建筑在水平面和垂直面上交替分层呈现，二者的交界面为构建新型空间创造了机会。

这个交界面体现了可持续发展策略——通过在翻修的公寓及外部屋顶空间的垂直和水平两个方向扩展绿色植物空间，最大限度地实现生态多样化和设计的可持续性。其成果"高空自然"是构筑自然的人工奇景，自然形态和自然现象营造出了奇妙而生机勃勃的景观。生态学与艺术在都市背景下相互碰撞，迸发出生命的活力。

交界面始于一个位于6m长的天窗下的室内花园，花园内种满了枝繁叶茂的大叶秋海棠和黑竹，这些植物在通往屋顶的悬空楼梯下方创造出一块不断向上蔓延生长的绿色挂毯。在雅致的竹叶上方，透过分隔花园和主浴室的玻璃隔断，可以看见一堵种满卫矛的绿墙。这些植物沿墙壁向上攀缘直至第二扇天窗，透过这扇天窗就能看见屋顶花园的各种植物。

高高的草丛环绕着屋顶花园，其间点缀着许多野花；它们柔嫩的叶子随风轻摆体现了风之花园独具一格的美丽。

悬于草丛之上的是双层错落式的露台。在较低的一层有一条砾石小径通向可观赏下东区景色的观景池，一间室外淋浴区，在楼梯隔断的另一侧还设有更为私密的按摩浴缸和阳光露台。五级阶梯通向带有室外厨房和烧烤休闲区的上层空间。桦树穿过这层的露台，在午后投下斑驳的阴影。在第二层还能看到女儿墙的绿色屋顶，苍翠繁茂的景天属植物创造出一种以天际线为背景的无边无际的绿色景观。

在女儿墙的对面，隔断向上延伸，营造出一个斜坡，人们可以躺在密植着植物的斜坡上看天边云卷云舒。穿过一条楼梯来到顶部，在那里向下俯瞰，可以将下东区的全景尽收眼底。

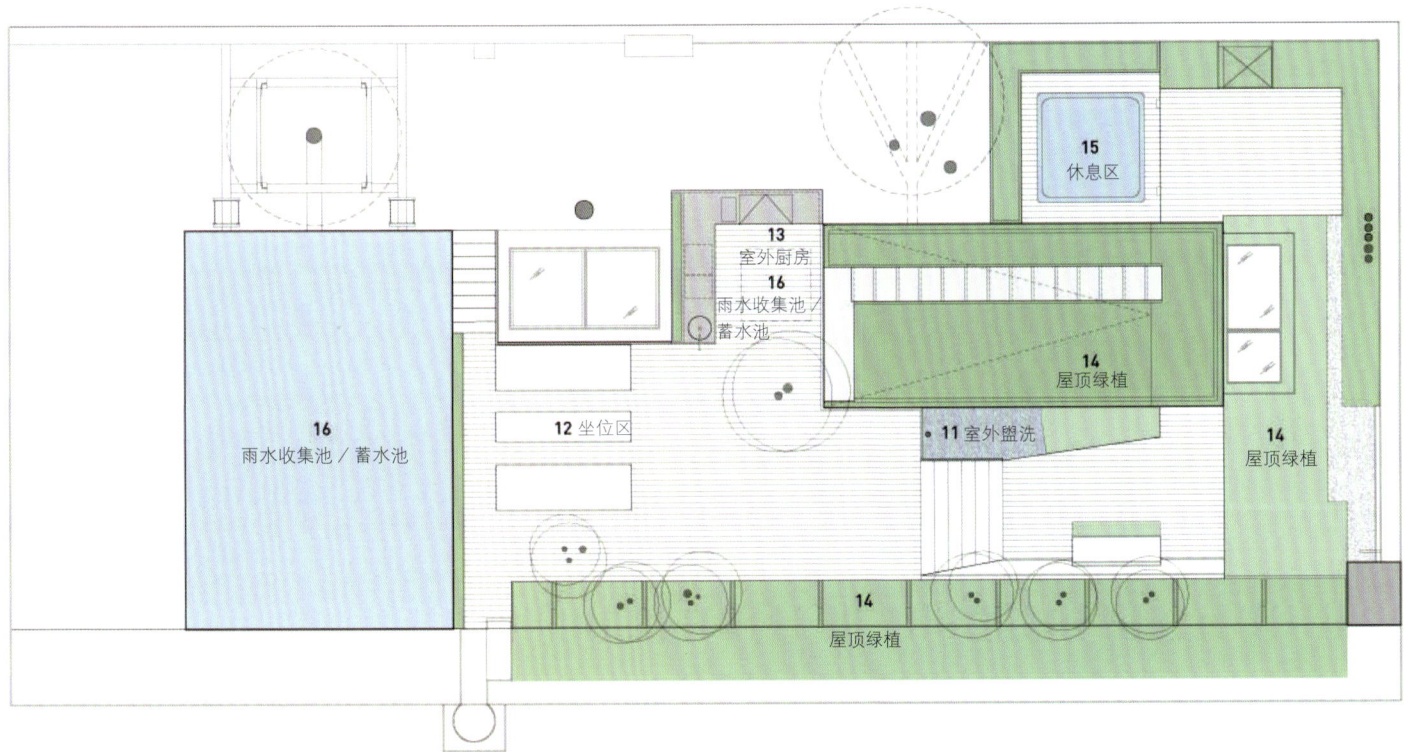

15 休息区

13 室外厨房

16 雨水收集池／蓄水池

14 屋顶绿植

12 坐位区

11 室外盥洗

14 屋顶绿植

16 雨水收集池／蓄水池

14 屋顶绿植

The project at 684 Broadway is an effort to explore the interface of the built and natural environment. Reconfiguring the space in between and making new connections creates more fluid passages; not blurring the line between landscape and architecture, but widening it. This thick interface creates the opportunity for new types of spaces. Alternating sheaves of landscape and build-ing on both horizontal and vertical planes create transitions within this widened line. It is a complex interface that is layered — the thicker the line the better — and results in a new spatial entity.

This interface becomes a sustainable strategy that aims to maximize biodiversity and sustainable design in this urban site by extending green space both horizontally and vertically within the renovated apartment and exterior roof space. The result, hypernature, is an artificial spectacle of constructed nature. Natural forms and phenomena are revealed and re-visioned into a magical landscape for living. Ecology and art meet at the surface creating an explosion of life within the urban context.

The interface begins with an interior garden beneath a twenty foot long skylight. Filled with large leaved Elephant

Ears and black bamboo, the plants create an ascending green carpet beneath the floating stairs to the roof. Above the delicate bamboo fronds, through a glass partition separating the garden from master bathroom, is visible a green wall planted with euonymus. This improbable swath of vertical vegetation climbs the wall colliding with a second skylight through which is visible the rooftop planting.

Suspended above the sea of grasses is a bi-level ipe deck. On the lower level a small gravel path leads to a lookout pod with views over the lower east side, an outdoor shower and on the opposite side of the stair bulkhead, a more private enclave with jacuzzi and sunning deck. Five steps lead to the upper level with an outdoor kitchen and grill lounging space. Birches punch through the deck at this level, creating dappled afternoon shade. From the second level the parapet green roof is visible. This swath of lush sedums creates an infinity edge of green with skyline as backdrop.

Opposite the parapet, the bulkhead rises into the sky. Densely planted with stepable plants one can lie on the slope and watch cloud rushing overhead. A staircase leads to the top from which there is a 360 degree view of the Lower East side.

240南中央公园

240 Central Park South

撰文：巴尔莫里建筑事务所　　图片提供：Mark Dye　　翻译：李沐菲

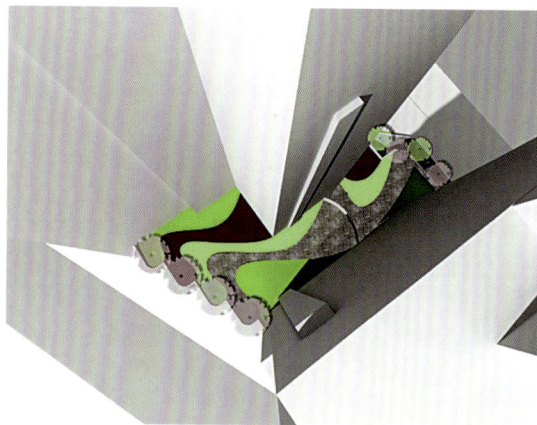

　　该项目建于 1940 年，由建筑师阿尔伯特·梅尔和朱利安·惠特西共同设计，是一座包含 325 个单元的双塔式公寓建筑，双塔之间由多个平台相连接。早在 70 年前进行设计构思时，绿色屋顶的设计就已被纳入总体规划之中。

　　该项目的绿色屋顶及入户庭院由巴尔莫里建筑事务所设计，通过建筑物本身及其屋顶来体现中央公园的特色。波状的灌木及景天属植物相互交织，边缘为石板组成的线条，宛如公园地面上突起的石块。

　　该项目的设计主旨便是展现多重的景观效果：行人可以看到矮墙上面露出的樱桃树，而居住于此的业主则置身于绿色、紫色和灰色的植物彩带中。从周围的建筑物和高层公寓的角度来看，这里错落有致的多层绿色屋顶仿佛又构成了一幅完整的景观。

Built in 1940 by architects Albert Mayer and Julian Whittlesey, 240 Central Park South is a 325-unit apartment building composed of two towers and a multitude of terraces linking the two at various levels. When conceived 70 years ago, the roofs were engineered to be planted.

The green roofs and entry courtyard of 240 Central Park South designed by Balmori Associates pull the character of Central Park through the building and up to the roof. Contoured ribbons of shrubs and sedums are interwoven with lines of slate quarried in New York State, mimicking the rock outcroppings in the park.

This landscape is designed to be experienced from multiple viewpoints: visitors walking by the building catch glimpses of the cherry trees peaking over the parapet wall, while tenants inside the building are surrounded by the rolling ribbons of green, purple, and gray. From the neighboring buildings and apartments above, the multiple levels of rooftops appear to join together into one unified landscape.

Villa

别墅／庄园

Manmade

木溪花园

Woody Creek Garden

撰文 / 图片提供：Design Workshop　　翻译：刘建明

科罗拉多州皮特金县的木溪花园已经融入了落基山独特的风景与生态体系，营造出一处集娱乐、休闲与冥想为一体的绝佳场所。场地层次的变化以及石头、光线和水元素的巧妙搭配，使得花园从原生植被到周围的白杨与针叶林之间的过渡十分自然。从设计伊始，建筑师就与景观设计师携手合作，令房屋内部与花园浑然一体。在周边自然环境所形成的特定风格下，木溪花园仍保持着整体项目最关键的元素——独特的地域感。

连接住宅的两个庭院使得每个房间都能欣赏到周围的景观。花园犹如住宅的绿色屋顶，使得下方的坡地不受暴雨、沙尘等的侵扰。水被作为一个统一的元素始终贯穿于设计之中，水雾、小溪、小瀑布以及宁静的池塘描绘出水的不同状态与形式，围墙外是保护完好的本土景观。

1　住宅和花园在坡度中完美结合
2　不同质地的材料构筑的花园地面

设计质量与实施

入口处有一处由围墙包围的花园，园内的花岗岩石板呈风轮状阵列排列，从别处移栽来的白杨林与之形成强烈的对比。花岗岩石板被当做长椅和桌子来使用，粗糙的科罗拉多沙岩墙和柱子与相互联接、精心设计的步行道又构成了截然不同的景观类型。由于直接暴露在水与空气之中，墙体自然风化形成的图案以及自然生长的青苔更加丰富了墙面的纹理结构。花园中央的巨石喷泉形成的水雾为这处封闭的空间平添了一种虚无缥缈的意境。

另一处花园与之相比却风格迥异，俨然是一处开阔、凸起的娱乐空间。设计师将此处设计成业主与朋友的聚会之所。由于此处午后降雨频繁，设计师有意将其设计成遮蔽式花园。由小瀑布组成的花岗岩墙构成了花园的背景幕布，瀑布水流猛然跌入宁静的水池里，此处的水池还可做温泉疗养之用。庭院的中央有一处独立的、呈几何形状的狭窄水池。从花园的凸起处观看，这处水池仿佛就是映照天空的一面镜子。周围的树林、建筑物、光线和阴影尽入水中，这种反射效果在无形中扩展了空间的广阔感。凸起的花园凌驾于周围景观之上，在这里可以领略到天际线边的辉煌景色。

水元素的设计运用贯穿于整个花园中。木溪花园的标志性特征就是从两处石墙之间的水源之地引流构建的独立喷泉。流水在沙岩石板上刻出来的浅水槽中流淌着，随即在光滑的黑色花岗岩上蔓延流淌。这一生动却并不复杂的流水效果在动静之间演绎得恰到好处——一边是光滑如镜的一池静水，另一边是水花四溅的喷泉以及遍布整个花园涌动的池水。汩汩冒泡的温泉就设在纯净无污染的池水旁边，池水清澈见底，池底光滑的黑色花岗岩石清晰可见。对水元素的千奇百怪的结构以及形态的娴熟运用是动感十足的该项目不可或缺的一部分，甚至在光线和明暗度的设计上也可以寻找到这一手法的蛛丝马迹。

1　入口景观
2　光滑如镜的石板
3　简洁大方的花园入口处
4　雕塑突出了乡土情结

环境敏感性与可持续性

　　用围墙来阻隔尘世的嚣扰，让山腹之地成为住家之所，这样的花园小区已和周围的自然环境融为一体。一系列可持续设计元素的应用使小区对环境的负面影响最小化。在该项目的植物设计中引入了亚高山地带的原生植物，一些商业苗圃中不常见的物种却可以在这里看到，由乔木、灌木、多年生植物构成的混合植被体系，包括白杨、云杉、橡树、北美黄松、卫矛、高山积雪、耧斗菜和樱桃树。该项目的设计理念就是要适应高海拔的自然条件，选用一些当地的种植原料和土壤，采用再种植的方法来建立本地原生植物群落。由于对白杨林的结构和生物学的深层次研究，该项目的植被得以在所处的自然环境中茁壮生长。成熟的白杨林形成的树冠不仅可以遮挡阳光、展示树影婆娑，同时也对生物多样性和生态复兴具有很大益处。该项目的植被体系还保留了一些与落基山地区生态群落截

然不同的视觉体验，其中包括空气中松树的香味以及耧斗菜花绽放时的粉白盛景。

　　绿色屋顶的设计不仅节约了能源，同时也将大部分热源隔绝掉。屋顶巧妙地利用了坡度，依据原有的地势、地貌创建了全新的室内外空间。屋顶和阳台的所有排水都被回收到干涸的井里，这样就不会增加径流的负荷。该项目严格按照当地政府制定的"可持续性开发指南"，并栽植了一些种类的植物来改善野生动植物栖息地的自然条件。

背景

　　该项目位于海拔约2700米的坡地上，针叶林与白杨林在此处混生。西部和北部的视野开阔；而其他方位的景色却隐匿在山谷当中。住宅和花园有围墙围护，以避免生态不稳定的地区受到外界侵扰。设计中考虑到了先前的气候研究结果，从而有效地阻隔了大风与

烈日的干扰。根据日光和气候而设计的可伸缩性阳台能使室内的布局根据每天的不同时段而进行相应调节。

　　木溪花园倡导住宅的统一和传统主题，但是更深层次的主题和设计目标却应该是与本土环境的和谐统一。明快而精细的几何元素构造出来的景观平面图将该项目融入到山谷当中，时尚的灯光和水元素融入了富于特色的建筑结构和空间。这是一个独具匠心、功能齐全的设计佳作，并与周围环境和谐互融，相生相息。

1　光滑如镜的石板与原生植被相得益彰
2　人工种植的白杨与周围的环境相生相息
3　花园的山地风景
4　步行道、楼梯和聚会场地

Woody Creek Garden in Pitkin County, Colorado creates a space for entertaining, relaxation and reflection that is integrated into the exceptional views and ecology of its Rocky Mountain setting. Variation in the quality and placement of stone, light and water allows a seamless transition from the garden's native plantings to the surrounding aspen and conifer groves. From the beginning of the design process, collaboration between the residential architect and Design Workshop landscape architect Richard Shaw blended the house interior with the garden. With a style defined by its natural environment, the garden maintains the overall project's most critical element: its sense of place.

Two courtyards interlink this single family residence, allowing each room to enjoy the visual landscape. The garden is built as a functioning green roof over a portion of the residence, leaving the steeply sloping site undisturbed. Water is the unifying element in the landscape design. Atmospheric mist, single rivulets, cascades and still pools portray water in its various states and forms. Just beyond the walls, the native landscape has been retained and preserved.

1　喷泉流淌出纤细的水柱

2　风化岩勾勒出特色水元素和天空的轮廓

3　室外壁炉

Quality of Design and Execution

A walled entry garden in a pinwheel arrangement of granite slabs on the horizontal plane is contrasted by naturalized groups of quaking aspens acting as vertical punctuation. The slabs function as sculptural seating benches and tables. Rough-hewn Colorado sandstone walls and columns contrast with the interlocking pieces of refined sandstone walkways. The textures of the wall stone are enriched by native lichen growth and patterns of natural streaking from water and air exposure. The garden surrounds a central stone misting fountain influencing the enclosed space with an intermittent ethereal ambiance.

The second garden, in comparison, is an expansive, promontory entertainment space. The design program anticipated large gatherings of the owners' friends. With frequent afternoon rains, the space is designed to include the possibility of tenting the entire yard. A cascading wall of water forms the backdrop to the garden, disappearing suddenly into quiet reflecting pools, one of which also functions as a recreational spa. A separate, geometric pool of shallow water is located in the center of the courtyard. From the promontory above, this plane provides a mirror to the dramatic ever-changing sky. This reflecting effect simultaneously expands a perception of spaciousness while drawing in imagery of surrounding trees, built structures, light and shadow. The promontory garden is elevated above the surrounding landscape to celebrate the dramatic views at the very edge of the skyline.

The element of water runs throughout Woody Creek

Garden. A signature feature of the garden is a single fountain that originates from its base between two rock walls. After tracking along a shallow trough carved into a horizontal sandstone slab, the water streams forth off the rough, chipped edge before bursting on a polished plane of black granite below. This dramatic yet simple effect plays on the contrast between the clarity of still or channeled water and the energy of water sprays and churning pools found throughout the garden. The bubbling spa is set next to a pristine sheet of water that runs transparent over smooth dark stone. This interplay between varied textures and energies in water is part of a greater dynamic in the garden, and it can be found in stone, light or darkness.

Environmental Sensitivity and Sustainability

By limiting the disturbance with walls and adapting the home to the hillside site, the residential garden becomes part of the natural setting. An array of sustainable design elements insure that impact to the environment is minimized. Plants native to the sub-alpine life zone are utilized in the landscape design. Often unavailable from commercial nurseries, some species were specifically grown for this site. A mix of trees, shrubs, perennials and grasses include: quaking aspen,

Colorado blue spruce, gambel oak, ponderosa pine, four winged suman, mountain snowberry, Colorado columbine and chokecherry. The concept of the garden is adapted to the conditions of high altitude with selections of plant materials, soils on the site, and revegetation methods to establish native plant communities. Close studies of the structures and biology of aspen forests allow the garden plantings to grow into the surrounding natural habitat. Maturing canopies of aspens not only provide shading for comfort and visual displays of shadow, but contribute to biodiversity and ecological renewal. Other sensory experiences distinct to the Rocky Mountains have been retained by this planting scheme, including the scent of pine in the air and the pink and white visual accent of the Columbine blossom.

The inclusion of a green roof creates energy savings for the residence and significantly shelters the heating required by the home. This roof elegantly utilizes the slope beneath it, creating outdoor and indoor space that is grounded in the original site topography. All drainage is captured from roofs and terraces and retained in dry wells on site, resulting in no net increase in run-off. Sustainable site development guidelines have been followed that were established by the local county and include enhancement of wildlife habitats

with the planting of certain plant species.

Context

The residence is located at an elevation of 9,000 feet on a sloping site of mixed conifers and aspen. Views are expansive to the west and north; other aspects are hidden in the hillside. The residence and garden are contained by retaining walls to avoid disturbance on an ecologically fragile site. A climate study was completed to design strategies to shelter outdoor spaces from the wind and capture sunlight. Access to sun and climate led to flexible terraces allowing furniture to be adjusted depending on the time of day.

Woody Creek Garden shares a unified, traditional theme with the house it is paired with. But the greater theme and design objective is a harmony with the high country environment. Geometric elements, both crisp and subtle, create view planes that extend out into the mountain air and horizon. Contemporary lighting and water displays are integrated into structures and spaces that are timeless in substance and appearance. This is a distinctive and functional work of design that engages and sustains its surroundings.

圣基尔根别墅

A Villa in St. Gilgen

撰文：Rainer Schmidt Landschafts Architekten　　图片提供：Raffaella Sirtoli Swantje Nowak　　翻译：王琳

在沃尔夫冈（Wolfgangsee）湖畔的圣基尔根村（St. Gilgen）有一处与新老社区相连的景观带在这里绵延伸展开来——平坦的草坪与峻峭的山峰形成了鲜明的对比，同时也为游人欣赏湖光景色提供了去处。

蜿蜒的小溪、玫瑰丛和一些多年生植物群落清晰地勾勒出了草地的轮廓；在这葱郁的草坪之上，考登钢的弧形镰刀状构筑物穿梭于此。镰刀状构筑物的颜色随着季节的变化而变化：湿润的季节里，它的色彩深沉而浓重；炎炎烈日中，则显现出明快的橘红色调；冬日里，白色的烟雾弥漫在其左右。

入口处花团锦簇，萱草和大丽花争相开放；池塘边，蓝色和红色的紫罗兰、萱草与蝴蝶花争奇斗艳。各种各样的草与井然有序的花相互搭配、相映成趣，为花园增色许多；紫杉篱与黄杨篱之间点缀着青青的草坪带。花园里，栽培的所有玫瑰都是名贵品种，生长在背阴和半阴凉处，时刻散发着浓郁香甜的气息。

花园里充斥着各种强烈鲜明的对比，如修剪整齐的草坪与颜色亮丽、粗放的多年生植物。每一个独立的休憩区设置也很有讲究且各不相同，如阳光普照的草坪、阴凉的树阴还有那散发着甜美香气的玫瑰丛。

An undulating landscape is the new connection between the new and old living quaters of a villa in St. Gilgen near Wolfgang Lake. The calm lawned area accentuates the huge alps and lake view panorama.

The modeled grass area is bordered by a small stream as well as roses and perennials. Corten steel arches like sickels cut through the green lawn. The sickels change their colour depending on the weather. The more humid weather makes them look almost black whilst in the sun they have a bright orange red glow. In winter they have white frost.

The entrance area flowers profusely with Dahlien, Hemerocallis. The pond is planted with blue and red violet hues with Hemerocallis and Iris. Different grasses add to the flower arrangement. A light colored grass band grows between the high yew hedge along the road and a low buxus hedge. All roses growing in shade and semi-shade aspects are historical species that have an intense scent.

The garden thrives on the strong contrasts such as the calm lawn and the wild and colorful perennial planting. Single seating areas are positioned with different characters depending on the daily preferences; Sunny overlooking the lawn, shaded under the trees or amongst the sweet smelling roses.

卵石滩别墅

Pebble Beach Villa Residence

撰文：Angela Watrous　　　图片提供：Tom Fox　　　翻译：刘丹春

　　该项目是在世界闻名的卵石滩高尔夫球场的 17 号球道上建造的一座面积为 929 ㎡ 的私人别墅。项目所处的场地面积为 6070.5 ㎡，全景视野开阔，从球场至太平洋尽收眼底。除私人别墅外，设计还包括了一片从别墅延展至球道的繁茂的草坪，形成了一幅绿色的水墨画。

　　该项目是一栋低矮的现代化建筑，具有石质覆面和铜质屋顶，且在建筑四周拥有广阔的自然景观和人工景观。该项目旨在建造一个简洁的极简抽象派环境，同时又具有自身魅力，既不会显得突兀又不会被周遭景观喧宾夺主。

　　这座别墅的石质窗户与传统窗户不同，设计师在墙上开凿超大尺寸的矩形洞口，并用钛包边，再将电镀玻璃安装在墙上，使别墅内部的人产生没有玻璃的错觉。设计师通过使用石灰石墙和花槽围墙与别墅相搭配的方式延展了室内与室外的视野；同时，花园空间将住宅与周遭景观相连接，营造出一处优雅的人工空间，使其与建筑完美地融于一体，既恰当得体又不受时间限制。

　　为了将水的使用量降低到最少，除了草坪之外，该项目中的植被均采用了既能抗干旱又能适应海岸气候的植物，其中包括常绿灌木草莓树、蒙特利柏树、紫

色的亚麻和薰衣草，并用野鸢尾属植物作为地被植物。开阔的庭院采用可渗水的风化花岗石制成，能够将暴雨的径流减到最小。

　　极简抽象派的艺术作品被精心地结合到该项目的设计中，进一步将自然元素与人工元素结合在一起。在庭院中，两个石质球体如同被随意投掷在风化花岗石上一般，这种排列方式使其与四周的树木和石灰石墙共同形成一种极简抽象主义的效果。

　　在前庭的第 18 号绿地上，设计师设计了一个极简抽象主义风格的雕塑喷泉。喷泉由底面积约为 0.65 平方米，高约 0.46 米的一块黑色花岗岩制成。喷泉的设

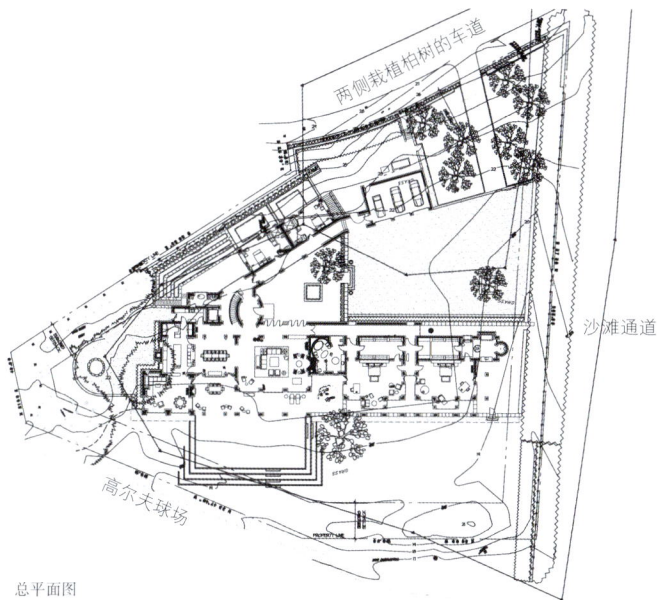
总平面图

计模仿了该项目地处陆地边缘的地理特点，将花岗岩沿对角切开，且一边稍高于另一边，水喷出来后，流过较低的半边，抽象地描绘了陆地与海洋会合的景象。

从露台处可以清楚地看到一个铁环，这是第三个雕塑，巧妙的设计使其看上去好像原本就存在的。铁环位于露台上的两个石柱所形成的框架中间，其形状增强了从别墅到海边的视野走廊，同时也不会阻碍或遮挡自然景观。

另外，设计师还设计了几座室外花园，用以突出别墅室内外不同的氛围。别墅的地下区域包括一个家庭剧院、一个运动室以及一个下沉式花园，花园中种满日本枫树，阳光透过枫树散落在石灰石阶梯上。位于两间主卧室之间的一处约 1.8 米 ×3 米的空间里密植了喜光的竹子，为人们提供了私密空间。两间客房也有各自的私人花园，其设计形式简洁，并带有修剪整齐的树篱，且树篱是一年一栽。为了将临近的公路与这些花园隔开，

设计师特别在石灰石墙壁上栽植了攀缘植物。

在厨房的外面，设计师设计了一个户外烹饪、就餐区域以及一个种满草本植物的厨房花园，内设烧烤工具和一个拥有最先进技术的披萨烤箱。原有的蒙特利柏树被移植到这个区域，形成一处屏障将其与邻近的住宅隔开，并使此处感觉更小、更私人化。当人们在室外就餐时，视线可经由柏树树冠，穿过树干之间，到达醒目的西部海岸线。

SWA performed full landscape architectural services for this 10,000-square-foot private residence on the 17th fairway of the world-renowned Pebble Beach Golf Course. The 1.5-acre lot has panoramic views across the fairway to the Pacific Ocean. The design included a lush lawn that flows seamlessly from the residence into the fairway, providing a unified wash of green all the way down to the cliffs' edge.

This villa is a low, modern building with stone cladding and a copper roof, surrounded by vast natural and manmade wonder. The goal of the landscape design was to create a simple, minimalist environment that was attractive without starkly standing out and taking away from the surroundings.

As opposed to traditional windows, this stone pavilion has oversized rectangular holes in the walls that are bordered in titanium; plate glass was then glued onto the walls, creating the illusion from inside that the building is an open-air structure without window panes. By using

site and planter walls made out of limestone to match the building, SWA extended this indoor/outdoor vision. The garden spaces integrate the home into the surrounding environment, creating an elegant man-made space that appears well integrated, contextually appropriate, and timeless.

With the exception of the lawn, water use was kept to a minimum by using drought-resistant, coastal-tolerant plantings, including Arbutus Marina evergreens, Monterey Cypress trees, purple flax and lavender, and Dietes as ground cover. The large auto court was created with permeable decomposed granite to minimize storm water run-off.

Integral to the site's design was the careful placement of minimalist artwork to further integrate the natural and constructed elements. On the auto court, two stone spheres appear to have been tossed onto the decomposed granite, using scaling to achieve a minimalist composition with the surrounding trees and limestone wall.

In the front courtyard on the 18th green, SWA designed

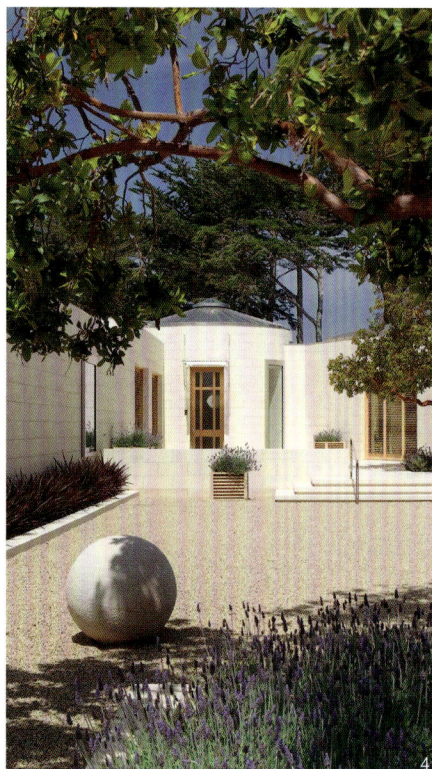

1、2　小型水景
3　下沉花园
4　别墅主入口处
5　场地原有的柏树
6　耐干旱的植物

a minimalist sculptural fountain out of a 7-square-foot, 18-inch-high block of black granite. Cleft diagonally, with one half wedged slightly higher, the fountain echoes the site's location at the edge of the continent. Water bubbles up and flows over the bottom half, creating an abstraction of land meeting ocean. A third sculpture can be best viewed from the terrace: a hoop so delicate it appears as if it only barely exists. Framed between two columns, the shape enhances the elegance of the view corridor down to the ocean without obstructing or interrupting the natural landscape.

In addition, SWA created several outdoor garden spaces to enhance the indoor/outdoor atmosphere of the residence. The home's basement area, which includes a home theater and an exercise room, has a sunken garden filled with Japanese maples that catch the light and bring it down

several tiers of limestone walks. Similarly, a 6 by 10 foot garden space between the two master suites overflows with light-attracting, privacy-providing bamboo. Both guest rooms also have their own private parterre gardens, simply designed with clipped hedges and annuals planted inside the hedged area. Shielding these gardens from the nearby road are backdrops of limestone wall covered with crawling vine.

Off the kitchen, SWA designed an outdoor cooking and dining area, as well as a kitchen garden with a vast array of herbs. Built-in barbecue facilities and a state-of-the-art woodfire pizza oven serve the space. Existing Monterey Cypress were replanted in this area as both a privacy screen from neighboring homes and a method of giving this enclave a smaller, more intimate scale. The outdoor dining views scan underneath the canopy of the Cypress, through

别墅的记忆 —— 布宜诺斯艾利斯私人花园

Memories of a Villa — A Private Garden in Buenos Aires

撰文：Jimena Martignoni　　图片提供：Facundo de Zuviria　　翻译：刘宏阳

1 漆成黑色的游泳池
2 天然山坡

该项目坐落在布宜诺斯艾利斯市北部的河滨区，地理位置得天独厚。这座花园二十年来一直是花园景观中的典范，设计师充分利用场地的地理优势，在浓郁的意大利风格中巧妙地穿插了具有阿根廷风情的景观元素，使花园设计别具一格。

设计主要基于以下两点考虑：其一是在设计私人花园时要特别考虑客户的个人品味和喜好；其二是要考虑场地内原有的景观元素。该项目的客户曾经在欧洲参观过一座意大利的别墅——Villa Lante in Bagnaia，那是他最钟爱的一座别墅，印象特别深刻；另外，该项目位于阿根廷 Plate 河畔陡峭的山坡上（Plate 河的拉丁语名称是 Rio de la Plata），视野开阔，风景宜人。设计师综合考虑了以上两点，因地制宜，在山坡的高处设计了很多几何形空间，这些几何形空间与别墅的住宅处于同一水平高度；而低处靠近河流的地区则采用自然、狂野的设计风格。

该项目规模宏大，景观错落分布，因此无法拍到别墅的全景照片。然而设计师巧妙的设计使众多景观元素有序地组合，将别墅中不同的空间连接起来。景观元素呈环形排列，在别墅中行走仿佛是进行一次新奇而神秘的旅行，但却有序可循。客户这样描述：在他记忆中的那座意大利的别墅里，每一个封闭的空间或是半封闭的空间都要由"门"而入。而在该项目中，"门"的形式更加多样化——可能是植物景观，可能是建筑元素，亦可能只是地势的高差。

简单来说，该项目创造性地采用了两种不同的景观：其一是 16 世纪的意大利风格的别墅景观，采用典型的几何设计和轴线设计；其二是热带花园景观，花园中水分充足、树木成荫，花园的造型突破了所有传统形式的束缚，在某种程度上颇有英式花园的风格。

这两种不同的景观都是根据场地中多样的地形进行设计的，并且尽可能地使花园与周边的连接形式多样化，设计突破常规，强调了花园的私密性。

设计体现了从开放性到私密性的转变、由传统花园到新式花园的飞跃，更体现了人造景观与自然景观的交融，这些都是花园设计内在价值的体现。然而，要达到所谓的转变、飞跃与交融并非易事。该项目设计成功的关键在于设计师因地制宜地利用原有的场地条件，想方设法地使花园景观多样化。客户本人也是一位景观设计师，通过适当地移土和填土，卓有成效地改变了部分场地的地表形态，创造出了客户向往的场地环境。

从该项目的布局上可以清晰地看到其与意大利别墅的联系：通过庄严的台阶、宽阔的露台和大气的阶地设计以调整场地的高差；两侧的大型植栽容器和雕塑大气恢弘；"环形旅行"的起点和终点以对称的造型设计清晰地标示出来；多处相似的绿色雕塑、意大利别墅中典型的喷泉、池塘和其他雕塑等。事实上，这

种联系在别墅的主入口处表现得尤为明显。当人们进入别墅或是离开别墅时，会发现别墅左侧有一座小院子，难以数计的仙人掌茂密地生长着，呈现出一派生机勃勃的景象，正是这处壮观的景象将人们引向花园的主入口，向上望去，一座半遮蔽的长廊将人们引向住宅。从这个角度看，花园是整座别墅设计中的重中之重，花园的地位要高于住宅本身。

花园中的植物枝繁叶茂、造型各异。植物的排列顺序十分讲究，前面植物的造型往往是后面植物造型的伏笔，使之在视觉上具有连贯性，同时增强花园的整体感。然而整个花园的背景与这些按照几何图形排列的植物造型完全不同。花园以高大的林木（如棕榈等）作为背景，形成一片茂密的树林，交织在不同风格的景观之中，巧妙地将不同的景观相互融合。这片树林异常繁茂，除了一些大型针叶树，还有蓝花楹、silk folk tree 和喇叭紫薇。一些林木是原本就生长在花园中的，另一些则是从其他地方移植来的。在炎热的季节里，树木展开如伞的华盖，树干也披上彩色的新衣，

有白色的、嫩黄的和粉色的等，花园中的花朵争相开放、万紫千红，仿佛要用亮丽的颜色来呼应炎热的天气。

花园的左侧建有砖石台阶，顺阶而下，可将花园中的植物景观尽收眼底，华盖如伞、枝叶葳蕤、层层叠叠，台阶两侧则生长着茂密的蕨类植物和大叶植物。花园中很多枝叶繁茂的植物也被保留下来，如竹子、棕榈、香蕉树、常青藤和很多蕨类植物。

花园中不同的景观营造出不同的氛围，这些景观的布局或依照时间顺序，或讲究空间顺序，同时也注重与人们的情感相呼应，景之所至，情之所至。

在别墅的最低处，一簇一簇的马蹄莲和其他草本植物相互簇拥着生长在山坡平坦的中心区，向山坡上

1、3 标示出道路的花坛
2 喷泉

望去，美丽的阶地和层叠的空间景观一览无余。砖石台阶在植物的簇拥中顺坡而下，花园的线条显得流畅而清晰。山坡上栽种了许多从别处移植而来的植物，这些植物已经在这里生根发芽、茁壮成长，非常自然地与原有植物交织在一起，毫无人造的痕迹。

这座花园注重"门"（通道）的设计，以花园中央的一处"门"为例：从空间上说，这座"门"酷似经过修整的灌木造型，此"门"一次只可允许一人通过。进"门"之后，随意设计的木质台阶将人们引向高处，继续奇特的环形之旅。这些木质台阶沿着陡峭的山坡而建，两侧棕榈成荫、热带植物郁郁葱葱，登至高处，放眼望去，河岸的景观壮丽秀美。

在欣赏了这片郁郁葱葱的热带景观之后，花园又展现出与众不同的半封闭景观。这处半封闭式场所适合举行正式的活动，中心是一座游泳池，四周设有凉棚、长椅和石质露台，顺着坡势连接着花园低处入口的景观区，这样花园的环形旅行就又回到了起点。人们可以在花园中自在地漫步——上坡下坡，走走停停，但永远有序可循，终点即是起点，一次又一次地体验无比奇妙的环形之旅。人们在花香四溢的花园中，或是

欣赏美丽的风景，或是驻足闲谈，亦或可以躺下小憩，逍遥自在。

多年以前，客户在层叠的景观之中建造了一座私人苗圃。由他亲自照看，时常到苗圃中劳作，其中摆放了许许多多的小罐子，这很可能是客户的无心之举，但却意外地创造了另一处景观：图形的重叠之美。尽管在项目初期设计师从遥远的南非、哥伦比亚和美国等地引进了很多植物品种，但现在花园中绝大部分的植物都是当地的品种，而这些异域的植物则孤零零地生长在山坡上。

该项目中植物繁茂、品种繁多，采光与遮阴的关系处理得非常巧妙，开放景观与封闭景观相辅相成、和谐交融，是花园景观设计中的典范。花园的景观造型新颖，更难能可贵的是，前后造型相互呼应、衔接紧密。当参观私人花园时，可能会好奇主人建造花园的初衷，可能会羡慕，可能会渴望，可能会把它刻在自己的记忆中，也可能希望重塑花园。但是我们绝不可能去改变，因为每一座私人花园在本质上都深刻反映着花园主人对生命的理解，对空间和自然的理解，这些都是不可改变的。

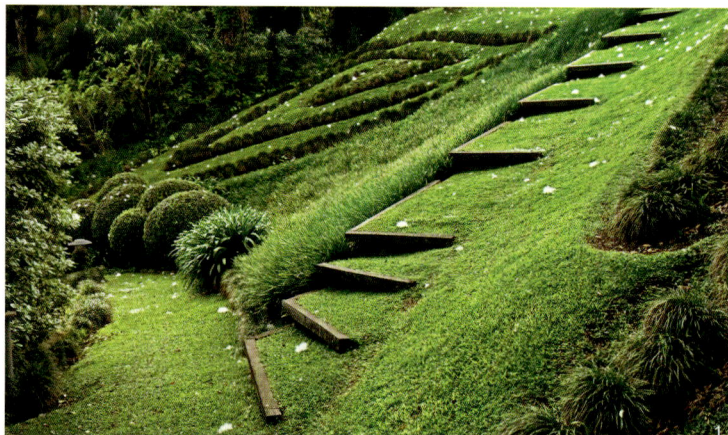

1 砖石台阶
2 热带景观
3、4 花坛
5 兰花

This very particular garden, located at the shore of the natural slope that face the river, north of Buenos Aires city, exhibits a landscape plan which has been modeled over almost twenty years and which combines the lines and shapes of a typical Italian villa with a very vernacular approach that refers to the original nature of the site.

This responds to two quite significant conditions. The first condition, especially relevant when talking about private projects is, precisely, the private preferences and personal taste of the owners and, the second, is the strong pre-existing elements of the landscape. In this case, the owner's favorite garden in Europe, which he had visited before starting the project, was the Villa Lante in Bagnaia, Italy; on the other hand, the site was a quite steep piece of land that dropped

away toward the coast of the Plate River, or Rio de la Plata, one of the widest in the world. As a response to this last situation and following the villa concepts admired by the owner, the project accommodates to the different "plateaus", offering more geometrical and formal spaces at the upper areas of the site, which coincide with the level of the house and the access street, and more natural and wilder spaces when getting to the areas at the bottom, closer to the river.

The scale is especially large. Although it is not possible to have one single picture of the entire garden, because of its size and propositions, what turns this into a series of connected diverse spaces is the decision of having it modeled in sequences. These sequences are related to the idea of a circuit, or journey, through very well differentiated areas,

which, according to the owner's description, are "rooms" or semi closed spaces that are entered through "doors". In turn, the doors, or accesses, are real frames materialized either with formal planting, with architectural elements, or with grading changes.

Basically, the garden takes and recreates two landscape styles: the geometrical and axial designs based on the iconic Italian villas of the 16th century, and the tropical garden, humid and shady, whose configuration is liberated from all formal clauses and which, in part, refers to the English garden manner. This double reference corresponds, first, to the naturally dynamic and versatile original topography of the site, and secondly, to the concept of differentiation between accesses areas, connected to the street and to the

public spaces of the compound, and those more private and less formal.

This emphasized progress from public to private, from formal to informal and from manmade to nature is probably one of the most inherent and primitive attributes of the garden design; however, it is not always possible to be achieved. The molding of the land, with the object of this differentiation, was a key action in this project. Based on the nature of the slopes, the owner—who happens to be the designer—did an important earth work, removing it at some spots and incorporating it at some others, to finally craft the diverse surfaces and ambiences he had originally envisioned.

The reference to the Villa Lante appears clear in the layout: the negotiation of the elevation changes with stately

stairs and large landing platforms and terraces, flanked by large plant containers and sculptures; the incorporation of symmetrical parterres which mark the beginning or the ending of an itinerary; the repetition of topiary design and the presence of focal points such as fountains, pools or other sculptural elements.

The main access to the house is, in fact, where this reference can be best appreciated. When entering the compound from the street, and after leaving, on the left, a yard where a massive group of pots with cacti and succulent species are picturesquely exhibited, the first image that is made out is a parterre composition which marks the main access to the garden. Before getting to this point, a semi roofed logia connects to the house; in this manner, the garden becomes a privileged space, more important than the building architecture.

The parterre composition turns into another framing element for a series of succulent plants exhibited on the ground, all of which have sculptural shapes that offer a subtle insinuation to what will be exposed later, when walking up and down the garden. However, what appears as a background, in deep contrast to this geometrical first welcoming layout, is a cluster of large trees and some palms that seem to establish the first dialogue between styles. Most of these trees have a stunning blooming: jacarandas, silk folk tree and trumpet tree combine with a few large conifers. Some of them existed at the site when the designer started to work on the project. When the warm season is announced on the air, many of these trees' branches get fully covered with a bunch of white, yellowish and pink orchids, as if the arrival of high temperatures could be only corresponded with the warm colors these flowers expose.

While going down the stairs and brick terraces that are developed on the left side of the garden, all these trees get visually incorporated into the landscape, conquering it with their crowns, their branches or only part of them. On the edges of these stairs, also appear shade species, ferns and large leaved plants which integrate with the original planting of the humid ravine such as bamboo, palms, banana tree, ivy and groundcovers.

The different achieved ambiences seem to follow one after the other, as spatial and emotional sequences.

1　蕨类植物和香蕉树
2　别墅的低处
3　雕塑
4　大型树木和兰花

At the lowest part of the site, from where the terraces and overlapping spaces can be fully appreciated, a cluster of calla lilies and other herbaceous occupy a flat central area. The sequential design of the green slopes is sometimes outlined with formal compositions and others with the combination of green edges and large steps. The natural sequence of time and the intentional play with time are easily perceived in this place. Mature species define the spaces but also adapt to them, and the already defined planes overlap almost naturally.

One of the "doors", or accesses, located at this flat area of the garden, is a geometrical topiary-like element whose central space is opened and has the exact necessary width to allow the passage of one person. Behind, an informal wooden long stair leads to the upper part of the garden, completing the circuit. This stair is developed along a quite steep part of the original slopes and among many palms and tropical species; once at the level of the house again, the river can be made out in the distance.

Right after these last tropical looking spaces the garden presents another semi closed formal area which hides the swimming pool. Pergolas, benches and a stone deck define the edges, which gently slope down to the first welcoming plane of the garden. The circular journey that is created through the walking experience of the garden is given by the possibility of sauntering around, descend or ascend and always going back to the same spots. This action can be only interrupted by other more passive practices typical of a garden: watching, smelling fragrances, resting, or lying down.

Some concealed spaces of the garden are reserved for a private nursery, plants growing and care, tasks that the owner decide to carry out on-site years ago. Dozens of small pots are grouped together and create, probably unintentionally, a common resource of beauty: the repetition of shapes. Although in the beginning the planting plan was not exclusively native, today most of the species are from local environments and regions. Many of the introduced plants were brought from far places, especially South Africa, Colombia and the United States; today these plants stand, stoically, on the slopes.

This site is, above all, an interesting combination of plants, light and shade, open and closed vistas, surprising shapes and successive allusions to other clearly recognizable places. A visitor could question the decisions and the results of a private project, could admire it or desire it, or could even attempt to recreate part of it, bringing it to his or her own memory, and reshaping it. What could not be done by a visitor is to imagine a private garden in any other way, correct it or modify it; because a private garden is the reflection of the owner's unarguable vision about life, death, places and nature.

4

慕尼黑博根豪森别墅花园

Villa Garden Bogenhausen, Munich

撰文 / 图片提供：Rainer Schmidt Landschaftsarchitekten　　翻译：马秀欢

别墅始建于 1923 年,具有典型的新古典主义风格,并于 1980 年进行了修缮改建。整个别墅花园通过简洁明快的线条和简练的建筑语言来展现其充满野趣的"自然"气息。各种高雅、优质材料的使用更加强化了这一设计理念,如自然石材以及青铜雕塑等。花园主要分为三个部分,另一部分是别墅东面的主入口,种植着经过修剪的黄杨和杜鹃花丛;第二部分是左右对称分布的花坛,花坛上植有经过修剪的黄杨,内部种植的各色鲜花体现了缤纷夏日这一主题;第三部分也是最大的一个部分,是一处倾斜的大面积草坪。位于草坪中轴线上的是跌水台阶,沿着台阶两侧摆放着种植于陶盆内的经过修剪的黄杨。在视线的水平方向上,人们可以看到跌水台阶的尽头有一个由常春藤修剪而成的沙发椅,背景则是有着不同高度、种类和颜色的修剪整齐的绿篱。在这片草坪的四周嵌有一道石墙,将整个草坪抬高,与周围环境形成高差。抬高的草坪、别致的绿篱、跌落的水景形成了一处与众不同的景观。在这个独特的空间中,鲜嫩的草地、夏季的风情、清凉的跌水……所有这些元素共同谱写了一曲美妙的交响乐。

总平面图

The original villa was built with a representative new classical style in 1923 and refurbished in 1980.Clear lines and a reduced architectural language dominate the garden character that is framed with a wild nature. Tasteful and high quality materials such as natural stone and bronze strengthen the clear concept. One of three areas is the main entrance east of the villa. The entrance is bordered by over sized box hedge and rhododendron. The second area is a box parterre right and left of the terrace planted with a summer theme. The third and largest area is a rising grassed dissected with cascading water steps that are framed with cut box hedges in terracotta pots. An ivy settee, with various trimmed hedges of different greens and textures as a background, forms the focal point at the end of the cascade. This lawn area is raised and supported by stone clad retaining walls. An unusual perspective is created by the graded grass, different hedges and the water feature the viewer. Together the elements rise and form a symphony of fresh green, summer flair and cooling waters.

1 花园东西的主入口处种植 着修剪的黄杨和杜鹃花丛
2 沿跌水台阶两侧摆放曾经修剪的黄杨
3 此处草坪是花园中面积最大的一部分，在跌水台阶的
   尽头可见一个由常春藤修剪而成的沙发椅

2

525,15
525,75
523,00
523.00
529.03

剖面图

360度全景别墅

A Villa with 360-degree View

撰文 / 图片提供：弗朗西斯景观设计公司　　翻译：张晶

人类居住地变迁，重新审视建筑环境与大自然之间的内在关系，营造舒适、健康、环保的居住环境已经成为未来城市发展的主流。环保建筑不仅专注于技术性问题，而且要真正实现人与自然的和谐统一。

景观设计师在设计中的主导思想始终如一，那就是将最能诠释自然风光的地域特征融入到景观设计中，旨在以最小的投入保留最真实的景观风貌，这个理念

在该项目中得到了完美体现。

该项目始建于 20 世纪 40 年代，占地 12 000m²，高居山顶，漫山遍野是有着 50 年树龄的挺拔松树，可以 360 度俯瞰贝鲁特城和地中海宜人的美景。

设计师亲手为这个古老的庄园打造了一座新型花园——小酒吧、水天相接的大水潭、玫瑰园、小型池塘、树阴下的休憩区和超大全景的视觉体验。小池塘

本身就是一个生态系统，是多种动植物的天然栖居地，户外空间各种配套设施更是一应俱全。

然而，打造一处美好的绿色家园不仅仅只是少利用、多保留，更重要的是要创造出适宜人居的环境。该项目中所有的软、硬质景观都遵循这种设计理念。设计师的首要意图是鼓励应用本地的软、硬质景观资源，如石头等花园的建筑主材料全部购于本地或是回

收再利用的资源，既环保又经济耐用，还可以减少生产新产品而产生的污染，因此在可持续景观设计中被广泛采用。

软景方面则选取适宜的环保型植物。适宜植物不仅维护成本低，而且不具有速生性或危害性；除了适应本地的气候条件外，还可节约灌溉用水。

花园里到处是香气袭人的薰衣草、茉莉、玫瑰、木犀花，以及长势喜人的月桂树、杏树、苹果树、柏树，这些具有可持续性的灌木和地被植物都被大量应用到了该项目中。此外，雪松、柏树、松树、橡树等多种树木还在小范围内进行了再培育，以保证这些现有物种的延续性。仙人掌被妥善栽种在角落里，而散落在私人休憩区旁的芳香玫瑰则给花园带来了别样的生机。为了节约灌溉用水，草坪的面积被尽可能的最小化。

在一步步营造更加舒适的可持续的人居环境的过程中，设计师一直与客户保持密切的联系，共同研究出满足一切合理要求的最佳设计方案。

The shifts of human settlements and the necessity of a change in attitude towards our built environment and its intrinsic relationship with its natural context, has become the ultimate new guide for people who want to live in comfortable, healthy, environmentally conscious homes. The environmentally-friendly architecture not only focuses on technological solutions, but also tries to reconcile man and nature in its formal dialogue.

As landscape architects, our major aspiration and commitment is first and foremost towards preserving a healthy ecological environment thus remodeling it in the most restorative and efficient approach. Whenever possible, our guide while designing is the nature of the terrain as it is best to employ and emphasize the nature's characteristics by integrating them in the landscape concept. The aim is to preserve the authentic appearance of the landscape with minimal input.

This was applied throughout designing the garden that spreads over 12,000 square meters located on the crest of a hill, with a dominating breathtaking 360-degree view of Beirut and the Mediterranean Sea. The wonderful thing about this house from the late forties with a hill rendered magical by 50-year-old pine trees is the creation of a young garden with an old soul.

The spiraling garden tempts visitors: a wet-and-dry bar, an infinity pool that seems to blend earth and sky, a rose garden, a pond, a shaded seating area, and of course, the larger than life panoramic view. The pond by itself acts like an ecosystem where flora and fauna profuse a natural habitat. This outdoor space has all the facilities and accommodations one may wish for.

However creating a good green home isn't just about conservation, about using less or saving more- although that's certainly a major part of it. It's about creating better homes that are easier on the environment, less expensive over the long term. All soft and hard landscape elements were therefore combined and installed in such way. As a first intent, we encouraged the use of local hard and soft landscape materials which is highly recommended as it will generate a detectable elimination of pollution. Local and recycled materials

such as clay, sand, wood, bamboo, recycled concrete are highly eco-friendly, durable and economic and therefore place major emphasis on sustainability. They reduce the pollution that results from manufacturing the new product. We abundantly incorporated the existing natural resources to build a green well structured garden in a very natural environment. Wood being ostensibly a renewable resource, it has been meticulously used for the flooring of this particular residence, as well as most of our projects depending on clients' requirements and allocated budget. Most of the materials used for the construction of this captivating garden are all local such as stone.

In terms of soft landscape, using the appropriate vegetation has its say in the environmental outcome. This is where and why we turn to adapted—also called native—plant species. Adapted plants are considered low maintenance.Aside from being adapted to the local climate, they require minimal or no irrigation but rather a very low maintenance such as mowing or chemical inputs such as fertilizers, pesticides or herbicides. They provide habitat value and promote biodiversity through avoidance of monoculture plantings.

The garden was therefore wrapped in aromatic lavender, jasmine, roses, cistus, glorious laurel trees, prunus, apple, juniperus. The abundant use of all these shrubs and ground covers were selected to be more sustainable. Furthermore, the rich existing vegetation was intentionally reproduced in smaller sizes to achieve a reforestation and ensure the continuity of the cedar, cypress, pine and oak species. Once the older trees have reached their maturity, the youngest ones would have blossomed. The cactus garden well niched in its corner adds an energetic feel besides the aromatic roses scattered next to the private seating area. Lawn areas were reduced to the minimum extent in order to reduce the high demand of water required.

When it comes down to taming the environment for greater comfort and sustainability, each member of our team in close and continuous coordination with the client will work together to come up with the most distinct solutions that tick all the right and required boxes.

山坡花园

Hillside Garden

撰文：安德鲁·阿博 安妮·卢修斯　　图片提供：凯斯·勒布朗　　翻译：张晶

该项目的设计目标是打造一处户外生活区，同时新建一间现代公寓作为老房子的扩建工程。设计师为这个面积不大的场地设计了一系列多功能的阶梯式景观空间。该项目的设计将场地陡峭的山坡地势发挥到了极致，后院至今保留了约9.14米的高度落差。大胆、结构化的景观设计将场地的建筑风格与住宅环境完美地融为一体。

设计师在补充与对比之间慎重斟酌，试图突出院落和建筑物的出众品质。景观设计也随之最大限度地利用场地，以求为客户提供丰富的户外活动空间。

原有的房屋前是一块平整的草坪，两侧对称铺设了小路并种植了绿树，附近的景观也都被其包围着。黑砖铺就的机动车道从房屋前一直延伸至房屋后，新

扩建的现代建筑和房屋后的各式景观隐约可见。

房屋的后部主要以直立锁边式覆铜板和原始楔形护墙板相间的结构为主。高层平台将扩建的房屋与新建公寓连接起来，成为最主要的户外生活区。平台使用蓝石条纹装饰，点缀其间的绿树和设计简单的草坪充分展现了现代建筑的神韵和直线条特色。

无论是从高层平台还是悬臂式的公寓露台上，人们都可以俯瞰到后山坡的全景。这里鲜明的建筑风格——薄石板饰面的围墙、钢筋框架和栏杆、木质阳台，与顺山坡而下的建筑景观自然整合，并与粗糙的填石路堤、茂盛的绿树形成了鲜明对比。

从高层平台或者房屋扩建的部分到公寓内较低的一层，都可以经由整体式的石质楼梯到达花园平台。

花园平台的铺装条纹与植栽的线形条纹保持一致，沿小路栽种的树木呈现蜿蜒之势。在一处低矮石墙的遮蔽之下，是通往观景台的狭窄小路和楼梯。观景台设置在高处，与山坡景观融为一体。木质和钢筋结构楼梯直接通往低层平台，通过林中小径和石质楼梯也可以来到这里。

低层平台有一块可供玩耍的草坪和一个小型篮球场，新种植的树木与下层林木投下的树阴起到了部分遮蔽作用。虽然这里是孩子们的主要游戏场所，但是整个山坡都可以让他们尽情玩乐，如并肩攀爬、在露台下玩耍、在树林中捉迷藏。

This project addresses the client's desire to create outdoor living areas in tandem with a new striking contemporary studio and architectural addition to a Colonial Revival home. The landscape contains a series of terraced spaces providing multiple experiences on a small site. This project maximizes the function of the steeply-sloping property containing an existing 30-foot elevation change in the rear yard. The design is a mediator integrating the site's architecture and residential context with a bold, textural and unexpected landscape.

Keith LeBlanc Landscape Architecture sought to emphasize the dramatic qualities of both the property and architecture through a measured balance of complement and contrast. The resultant landscape maximizes its site while providing a variety of outdoor spaces for both the clients and their three young children.

A taut lawn panel framed by walks and planting serves as a foreground to the symmetrical facade of the existing house and embraces the neighborhood context. A driveway paved in black brick extends past the house front and provides a glimpse of the contemporary additions and the diverse qualities of the landscape beyond.

The rear elevation of the house is dominated by standing-seam copper clad additions juxtaposed with the original clapboard structure. The Raised Terrace links the additions with the new studio creating the primary outdoor living space. This terrace is comprised of bluestone bands that

complement the texture and linearity of the contemporary architecture framed by simple panels of groundcover punctuated by trees.

From the Raised Terrace, as well as from the cantilevered studio decks, one overlooks a diverse sequence of spaces that step down the rear slope. The crisp vocabulary of the architecture: stone veneer walls, steel framing and rails, and wooden decks, weaves itself into the built landscape down the hillside. These architectural elements contrast with rough stone embankments and lush plantings that become increasingly less structured as one descends.

The Garden Terrace is accessed from the Raised Terrace by a monolithic stone stairway, as well as from the interior lower levels of the addition and studio. The paving bands of the Garden Terrace give way to linear bands of planting, and mulch paths meander through drifts of textured planting. A low stone wall conceals a narrow walk and stair that descends to a Viewing Deck which is at once perched above and integrated with the hillside. A stairway of similar wood and steel provides direct access to the Lower Terrace while a woodland walk of mulch paths and rustic stone stairs provides an alternate route.

The Lower Terrace includes a play lawn as well as a small basketball court tucked to one side and partially screened by the canopy of new woodland trees and understory plantings. While this is the primary play space for the family's three young children, the entire slope was conceived to provide opportunities for play: boulders to climb upon, a deck to play beneath, a woodland to hide within.

石间农庄

3 Pedras Farm

撰文：Isay Weinfeld　　图片提供：Cristiano Mascaro

　　该项目占地面积为 25 万平方米，对设计师来说这是一项非常特殊的任务，因为它与 19 世纪的建筑有着历史性的关联。主体建筑建于 1871 年，设计师对其周围曾是奴隶和看门人的居所以及仓库也进行了改建，并添加了很多现代的创新性元素。

　　尽管几年前曾对其进行过修缮，但仍缺少很多基础设施，如公共卫生间等，而且也不能满足客户在周末与亲朋好友欢聚一堂或退休后颐养天年的需求。为了改善这一现状，设计师将曾经是看门人居住的房屋改建成健身房，将奴隶居住的房屋改建成会客室，而将旧仓库改建成小教堂。

　　设计中最大的挑战是如何通过最小的改造使其既符合现代生活的要求，同时又尊重建筑的历史意义。根据这样的设计理念，设计师设计了几个盒式房间来满足现代生活的需要——浴室和储藏室设计在镀铜的矩形空间中，楼梯设置在皮质外壳的矩形空间中。这些矩形空间独立建造，高度低于原有建筑，为人们提供了一个欣赏原有建筑的理想视角。

　　树林旁还建造了一处供人们休憩的亭子和游泳池，风格简约、朴素，完全不同于其周围那些建于殖民时期的华丽建筑。

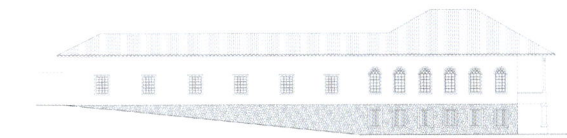

主体建筑立面图

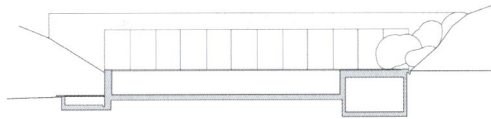

亭子东立面图

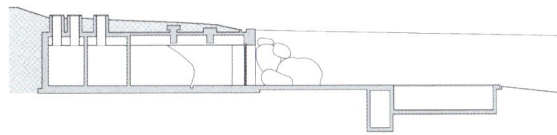

亭子立面图

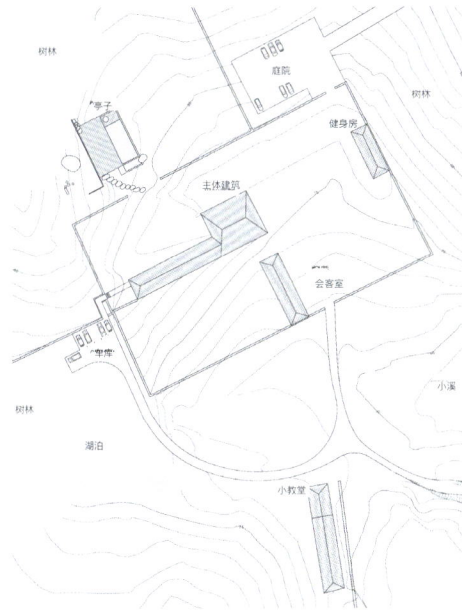

总平面图

The project for 3 Pedras farm was for us a very peculiar assignment as it implied interfering in 19th-century buildings of historical relevancy.

Located in a property ca. 25 ha in size, the main house, built in 1871, and its surrounding constructions for slaves and caretaker's lodging and storage were all to be adapted to modern or new uses under rigorous observation of heritage-protection laws.

Despite some previous remodels carried out over the years, the house lacked very basic facilities – such as sanitary ones – and did not suit the client's needs as a place to gather family and friends over the weekend or, eventually, to live in after retirement.

The biggest challenge was, of course, to make it fit for "modern living" with the least of interventions – least but not "invisible" as in order to respect the historicity of the building it was imperative to honestly and distinctively show old from new, and to avoid mimesis and blending by all means.

Solution came with the creation of boxlike containers to accommodate the new demands: bathrooms and closets in copper-covered boxes; stairs in a leather-covered case. Built independent of the existing structure and standing much lower than the ceiling of the house, these receptacles allow for a clear and fair perception of the original construction.

Also made suitable to hold modern comforts were the once caretakers' domicile—turned into a fitness room; the former slaves' lodge—turned into guest rooms; and the old deposit—converted into a small chapel.

Aside, by the woods, a leisure pavilion and a swimming pool were built, simple and discreet not to compete with the splendid colonial architecture of those pre-existing constructions.

巴西自然景观与人造景观的交融

Connecting Natural and Manmade Landscapes in Brazil

撰文：Jimena Martignoni　　图片提供：Chacel Office　Jimena Martignoni　　翻译：刘丹春

占地 17 000 万平方米的一排排无边际的橘树形成了雄奇瑰丽的人文景观，巴西景观设计师费尔南多·夏赛尔（Fernando Chacel）在这里建造了他最新的项目之一——位于圣保罗州农业区的一个私人 fazenda（葡萄牙语"庄园"之意），连绵起伏的、肥沃的绿色田野是这里最主要的视觉标志。

这个令人赞叹的人造景观被另一个更为壮观的自然景观所围绕，即大西洋沿海森林区（或称大西洋雨林）。作为巴西农业和工业生产之源，这个森林体系囊括了很多当地的植物群落，其中大部分是特有物种（只生长于这个地区的物种）。

当第一批葡萄牙殖民者来到巴西的时候，大西洋沿海森林区覆盖了整个国家 12% 的土地，与亚马逊雨林一起并称为南美大陆上最大、最重要的森林体系。由于人类活动的影响而导致森林长期以来急剧减少，为了保护该地区，1992 年联合国教科文组织宣布将其划为世界生物圈保护区。尽管该地区的面积在减小，但由于海拔变化而引起的气候、植被、日照时间的变化，使得该地区的生物多样性能够保持在一个很高的水平，该地区仅 10 000 ㎡ 土地上拥有的植物种类就比其他任何一个地方都要丰富。

20 世纪 30 年代，世界经济大幅下滑之后，咖啡种植园通过种植橘树来增加财政收入，因而这片农业区也宣布成为大西洋沿海森林区的一部分。在这独一无二的自然景观之中，业主们甄选出 25 万平方米的土地——在这一自然区内，墨西哥建筑师 Ricardo Legorreta 设计了一座占地面积约 4000 平方米的主楼

1　主楼与当地植物

2　橘树种植园鸟瞰图

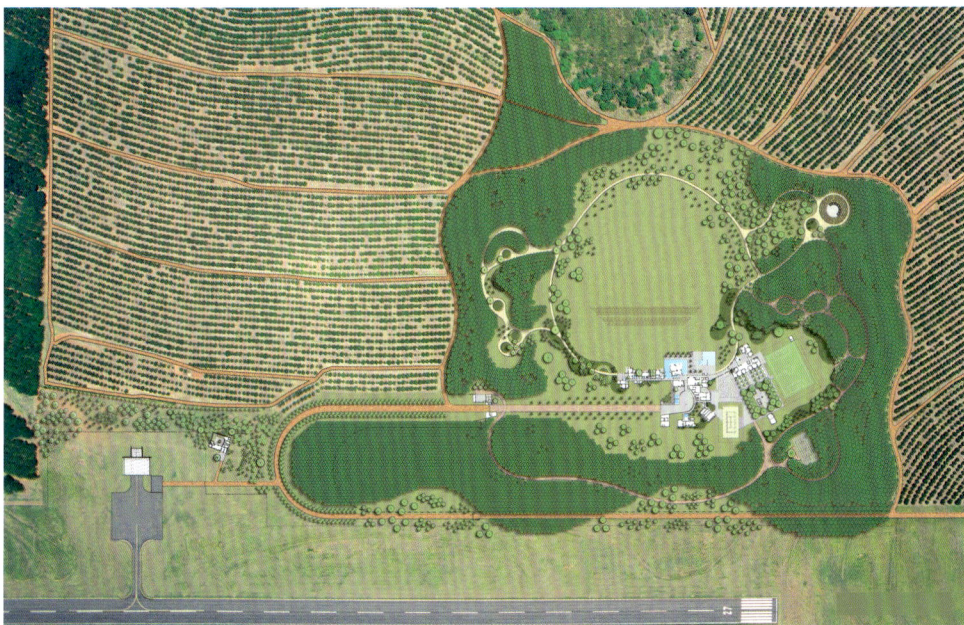

总平面图

和总占地面积 2.4 万平方米的庭院、运动场地和游泳区域。

2003 年，当费尔南多·夏赛尔首次受托设计主楼及其周边的过渡景观时，他不得不面对这个项目极其宏大的规模和紧密相关的视觉效果和文化因素。

在职业生涯伊始，夏赛尔曾与罗伯特·布雷·马克斯共事，前者的设计风格和观点在很大程度上受到了马克斯这位著名的景观设计师和艺术家的影响。另外，曾与布雷·马克斯携手工作过的巴西植物学家 Mello Barreto，对夏赛尔的职业生涯和风格形成也产生了重要的影响。在他们共同实施过的项目的基础上，夏赛尔发展了他称之为"生态进化"的设计过程——由于无限制的城市化以及人类活动，自然生态系统严重扭曲，"生态进化"是以更新自然生态系统为目的的人为过程。这个过程主要包括导入当地的植物种群和群落，重建自然栖息地和濒临灭绝的植物群落，并通过生态演替恢复它们之间的生物学联系（生态演替是指在某个特定区域内的植物群落随着时间的缓慢推移而转变成新的植物群落的生态过程）。在自然系统中，恢复当地的植被群落有助于保护当地的动物种群。

对于这个项目，费尔南多·夏赛尔认为让建筑与大自然巧妙地融合起来是设计的主要目标，这种融合将随着时间的推移而逐渐完善。

规划之初首先需要解决的问题之一就是改变该地区由植被环绕而成的形状。这块场地属于橘树种植园的一部分，是一个几乎完美的边长为 460m 的正方形，但它的几何边界与景观的自然轮廓并不吻合。因此，设计师决定采用有机的方式来界定这个地区的边界，即设计一个与天然坡度相协调的新的曲线边界并划定一条内环路线。

很早之前，这个地区与大西洋沿海森林区仅在项目的西面相连接，彼此之间没有缓冲和过渡区域。为此，夏赛尔设计了一个附加的毗连区域，并采用大西洋沿海森林区的物种进行重新造林，这样既避免了两个地区继续分隔又促进了交互式的自然进程。

为了该地区的概念规划夏赛尔设计了三种不同的

景观格局：森林（bosque florestal）、公园（parque）、建筑庭园（jardins de pre-arquitectura）。通过将这三种迥然不同的景观形式整合在一起，表现出融合区域间的不同之处，这个区域包括当地的热带森林和广袤的绿色种植园以及人文庭院和主楼四周的小径。

森林中的植物均为当地物种，这是建立在地域差异性的基础上，并且依据植物学知识和当地文化进行选择。根据"生态进化"过程，首先种植的是"先行"树种（生长较快的树种），以便在较短时间内形成较大的树群，并提供更大的遮阴面积；另一方面，"二批"树种生长缓慢，自然总量也少。

森林中一些最重要的树种有粉花风铃木和其他风铃木属树种，以及苏木、合法卡林玉蕊木和军刀豆木。通过种植大西洋沿海森林区的所有典型树种并利用生态演替的原则，可以使"森林"与自然植物群落结构相匹配；然而，种植设计并不是为了重造一个森林，而是使它具有森林的功效，可以为人类所用。为此，种植结构被设计成边长为3.6m的方格状，不同物种之间留有更大的空间以方便人们行走。

该地区的中心（新建环路的中心部分）与橘树种植园的主楼相对，主楼外围大约5万平方米的区域是夏赛尔设计的第二种景观格局——公园。该区域是被作为大型的中央绿色景观延伸区来设计的，其中只增加了一条从主楼延伸到周围田地间的200m长的花岗岩石阶路。

公园的树木选择更广泛，包括装饰物种和移植到圣保罗州的非当地物种。为了突显出区域间的界限，设计师们在公园边上种植了多种棕榈树，例如大王椰子（Roystoneas regya）、皇后葵、亚力山大椰子（Archontophoenix alexandrae）和菜王棕。

建筑庭院（或称做 jardins de pre-arquitectura）——将庭院、小径和主楼附近的空间架构起来，并将当地的热带花草和灌木与其他外来物种相结合：本土植物包括凤梨属植物、喜林芋属植物、海里康属植物和紫山姜（Alpinia purpurata），以及一些外来植物，如白色天堂鸟（Sterlitzia）、萱草属植物（Hemerocallis）和鳞芹属植物（Bulbine）。最成功的设计是花床与灌木的色彩搭配，同时也强调了典型的雷可瑞塔风格住宅的浓烈

色调。通过这些设计，建筑庭院为建筑和自然景物之间提供了主要的视觉联系。

为了与庄园相连接，农场在一排排无边际的橘树之间修建了一条蜿蜒的通道，在临近主楼的位置，这条通道转变为与私人机场跑道相平行的直线形车道。为了遮蔽私人跑道，设计师们在跑道两侧密集种植了不扩散生长的竹子。刚种的时候竹子只有约2米高，并没有达到设计师们预期的效果；但是在三年后的今天，生长过高的竹子不得不垂下身来形成一个苍翠繁茂的天然遮篷。同样，不同的当地物种被种植在森林边上用以突显边界，现在这些当地物种繁茂生长，并达到了连接景观的目的。

在这个项目中，植被成为了主要的建筑元素。这一方法已经由布雷·马克斯在巴西进行过深入的探究，并且成为了一个标志性的景观概念，这一概念不仅启发了很多拉丁美洲的设计师，而且世界上其他地方的设计师也广受影响。夏赛尔便利用他对当地植物物种知识的掌握以及生物学行为的了解重建了巴西最重要的自然景观之一。

17,000 hectares of never ending rows of orange trees shape the imposing cultural landscape where Brazilian landscape architect Fernando Chacel created one of his most recent projects: a private fazenda—Portuguese for hacienda—situated in the farming lands of the state of São Paulo, where productive undulating green fields become the major visual trademark.

This impressive manmade landscape is surrounded by an even more striking natural formation, the Mata Atlântica or Atlantic Forest. Source of the agricultural and industrial production of Brazil, this formation houses large local plant communities, most of which are endemic species (species that grow only on this part of the planet).

When first Portuguese colonizers arrived in Brazil, the Mata Atlântica covered 12% of the total country's area, and together with the Amazonian were the largest and most important forests in the continent. In 1992, after being drastically reduced over time, as a consequence of human activity, the area was declared a World Biosphere Reserve by UNESCO, for conservation purposes. In spite of the reduction of this area and due to variations in altitude and consequently in climate, vegetation and sun-exposure, the level of biodiversity keeps really high, containing in just one hectare more trees' species than any other part of the world.

This farmland had claimed part of the Mata Atlântica in the 1930s, when coffee plantations added orange groves to broaden their financial base after the worldwide economic crash. Enclosed in this singular natural setting, the owners chose a 25 hectare-central spot where Mexican architect Ricardo Legorreta designed a main house of almost 4,000 m² and a series of courtyards, sports and pool areas that together cover a total of 2.40 hectares.

In 2003, when first commissioned for the design of the "connecting" landscape of the house and the enclosing natural setting, Fernando Chacel had to face the especially massive scale of the project and the combination of those highly relevant visual and cultural existing conditions.

Fernando Chacel is one of the most recognized landscape architects of Brazil, with offices in Rio de Janeiro and São Paulo. Sidney Linhares, his partner in the São Paulo office, is in charge of all local projects, including the fazenda.

Having worked with Roberto Burle Marx at the beginning of his career, Chacel's designs and vision have been highly influenced by this famous landscape designer and artist. On the other hand, Brazilian botanist Mello Barreto, a specialist who used to work hand in hand with Burle Marx, was another important reference for Chacel's professional career and profile modeling. Based on the work they carried out together, Chacel developed what he named the ecogenesis process: a manmade process that seeks to regenerate natural ecosystems which have been deeply modified as a consequence of unlimited urban growth and development of human settlements and activities. This process primarily consists of reintroducing native flora and plant communities, re-establishing natural habitats and extirpated plant communities, and restoring their biological associations through ecological succession. The restoration of natural ecosystems with native plants helps also to protect remnants of local fauna. (Succession is the gradual change of dominant plants in a given area over time; this transition eventually creates new communities).

For this project, Fernando Chacel understood that the subtle integration of architecture and nature would be the main objective of his work, and one which would develop over time.

One of the first things that Chacel decided when started the planning process, was to modify the shape of the site that had been cleared of all vegetation. This piece of land, that was part of the fazenda's orange plantation, was an almost perfect square of 460 x 460 m whose geometrical boundaries did not match the naturalistic contours of the landscape. Therefore, he decided to redefine the site's perimeter with a more organic approach and designed a new curvilinear edge that follows the natural slopes and delineates the course of an internal loop road.

Before, the site and the Mata were physically connected at only one point, on the west side of the project, with no buffer or transition between them. For this reason, Chacel designed an additional contiguous fraction to be reforested with species of the Mata, thus avoiding further fragmentation and encouraging interactive natural processes.

For the conceptual layout of the site he created three different landscape patterns: the bosque florestal or forest; the parque or park; and the jardins de pre-arquitectura or architectural gardens. Through the design of these different but integrative areas, the landscape plan seeks to gradually mingle the different scales, from the regional tropical forest and the vast green plantations to the human-scaled courtyards and paths around the house.

The plant selection was also based on the differentiation of the areas: for the forest, all species are native and were chosen based on their local importance, botanically and

1 住宅周边的花园和当地的棕榈树
2 平台屋顶
3 主楼与庭园前的石阶

culturally. Following the ecogenesis process, the "pioneer" species (those species which grow faster) are planted firstly in order to generate larger groups in less time and to provide larger shady spots; the "secondary species", on the other hand, grow much slower and the total number is consequently lower.

Some of the most important species planted in the so-called forest are Tabebuia roseo-alba and other Tabebuia species, Hymenaea stilbocarpa, Cariniana legalis and Machaerium villosum. The list of plants was selected from a scientific research conducted in 1997 by Adriana de Fatima Rozza, which provided specific flora inventories for the Mata. By planting all the typical species of the Mata Atlântica and utilizing principles of ecological succession, the forest seeks to match this natural structure; however, the planting plan's goal is not to recreate it as if it was the original one but to intend a space for human use too. For this reason, the planting configuration is a 3.60 m × 3.60 m grid that generates wider spaces between the species and allows people to walk.

At the center of the site (the inner part of the newly created loop) is located the house which faces the orange plantation; the immediate surrounding area, of approximately 5 hectares, is the second landscape pattern created by Chacel, called the park. Designed as a large central green extension, this area is only interrupted by a number of 200 meter-long granite steps which descend from the house to the surrounding fields.

The plant selection for the park is more open and includes ornamental species and some non-natives that have been naturalized in the state of São Paulo. On the sides of the park, in order to demarcate different areas, the designers planted diverse native and non-native palms such as Roystoneas regya (royal palm), Syagrus romanzzofiana, Archontophoenix alexandrae (Alexandra palm) and Roystoneas oleracea.

The architectural gardens, or jardins de pre-arquitectura, frame the courtyards, paths and spaces right next to the house and combine native tropical flowers and shrubs with other introduced species. Some of the native include Bromeliads, Philodendrons, Heliconias and Alpinia purpurata (red ginger) and some of the exotic plants are Sterlitzia (white bird of paradise), Hemerocallis (day lily) and Bulbine (burn jelly plant). The most well-achieved design effects are the ones created with flowering beds and shrubs whose colors match and emphasize the strong color palette of the residence, typical of Legorreta's architecture. In this manner the gardens provide a first visual connection between the buildings and the site.

To reach the fazenda, the farm compound presents a winding main access road that goes between infinite rows of orange trees and which, when coming closer to the house, turns into a linear entrance driveway that runs parallel to the private runway. In order to screen the latter, the designers planted a thick double allée of noninvasive bamboo; when initially planted, the bamboo was only about 2 meters tall and didn't really achieved the intended effect of the designers, but today, after three years, the plants are so tall they bend over and create a lush natural canopy. In the same way, the different native species that were planted to restore and to recreate the edges of the forest now look luxuriant and begin to achieve the main goals of landscape reconnection set by the conceptual layout.

In this project, vegetation is clearly used as an architectural element. This is something that has already been explored in Brazil by Burle Marx, and which became an iconic landscape concept that inspired other designers, not only in Latin America but in the rest of the world. Fernando Chacel also used his knowledge of local species and their biological behaviors to partly recreate and partly restore one of the most important natural landscapes of Brazil.

人工河流景观

Manmade Riverscape

撰文：狄恩·海罗德　　　图片提供：Danny Kildare　　　翻译：谷晓瑞

1　傍晚瀑布景观

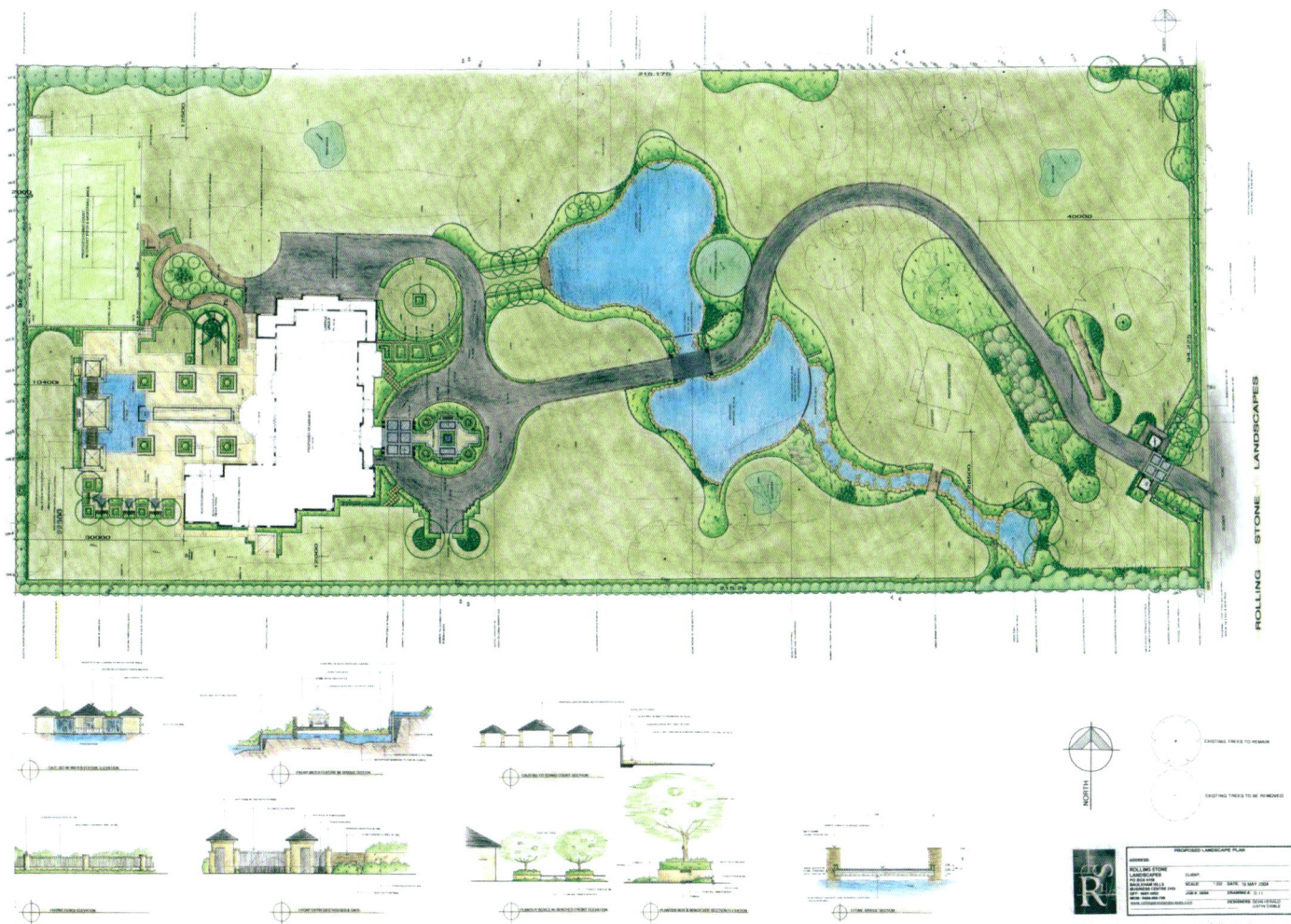

总平面图

要设计改建一座占地面积为 20 234 m² 且布满灌木丛的牧场，可能很多人都会知难而退。但对于来自悉尼的获奖景观设计师狄恩·海罗德来说，设计一些独特的景观将是一次难得的挑战。

由于狄恩的公司——Rolling Stone Landscapes，在项目初期便参与其中，这使得他对一些关键元素（如房屋的位置）能够提出可行性的建议。

"这个地方原来只有一个小池塘和一片茂盛的灌木丛，"海罗德说，"池塘的地理位置非常好，整个牧场的排水都可以聚集到那里。因此，我们将房屋所处的位置抬高，这样雨水就可以顺着屋顶流下来进入后期

要建的水库里。"房屋的位置要很靠后，这样可以设计一个气派的入口；一条蜿蜒的车道向内延伸，车道两侧是修剪整齐的草坪和花坛。另外，较高的地势也可以提供良好的观赏视角。

确定房屋的位置后，接下来要规划水库的位置和规模。"水库的容量要足够大，可以容纳流经 1000 m² 屋顶和 500 m² 地面的水流，这些汇集起来的雨水可以用来回收灌溉。"规划中两座水库总容积不少于 2 百万升。两座水库上都有瀑布，由一条人工河流连通。"将水从水库下游抽到上游，经过岩石和瀑布时进行充气，以保证瀑布的效果。"

215

设计师在项目初期就参与其中的另一个好处就是可以进一步设计天然的洼地，将所有从草坪、车道、路面汇集的雨水引入水库。这些回收的雨水可以用来灌溉草坪和植物，形成一个自给的园林系统。

设计的目标是依据庄园的需要建造天然景观。尽管选用的植被不全是本土植被，但园林绿化也要突出自然的主题。

"一些天然的植物，如洼地里的狐尾草在微风中摇摆呈现出动感之美，"海罗德说，"我们还在叶子柔软的栀子花丛中添加了像兰花和新西兰亚麻这样的草本植被，使植物搭配保持平衡。"

该项目中选用的植被还有修剪过的黄杨树、绿篱以及樱桃树和枫树这样的观赏树木。植被不断生长，点缀在房屋周围，使整体效果更加完美。

场地里还有一个精修的四洞高尔夫球场。球场上的每个球穴都经过精心设计，不但设有沙坑，还有带有标杆的水面，均体现出了庄园的特色。

Being faced with a scrub-covered 5 acre paddock can scare off most people, but for award winning Sydney based landscape designer Dean Herald, it presented a challenge to create something special he could just not pass up.

Dean's company, Rolling Stone Landscapes, was engaged at a very early stage in the project which meant that Dean could advise on key elements, such as the position of the house.

"All that existed on the site was a small pond and overgrown scrub," says Herald. "The pond was located in a perfect position to collect runoff from the entire property so we decided to position the house so that it was high enough to provide good fall so that the rainwater collected from the roof could be used to fill the future dams." The house was also set towards the rear of the property to help to create a grand entrance with a long meandering driveway through manicured lawns and gardens. In addition, an elevated position would provide the best outlook.

After positioning the house the next priority was to design the location and size of the dams. "The dams had to

be large enough to store 1000m^2 of roof and over 500m^2 of paving runoff, both of which would collect rainwater that is recycled for irrigation." Two dams were designed with a combined capacity of over 2 million litres. Added to the dams were two waterfalls linked by a man-made stream. "Water is pumped from the lower dam to the top of the stream, as it passes over the rocks and various waterfalls it is aerated ensuring the quality is always maintained at a high level."

Another benefit of being involved at such an early stage of the project was that Dean and his team could further sculpt the already existing natural swales to direct all the collected water from lawns, driveway and paving into the dams. The water is then re-used to irrigate the lawn and plants creating a self sufficient landscape.

The aim was to create a natural landscape tailored to the properties needs. Natural sandstone from the site was re-used along with imported local sandstone to line the dams and stream in organic shapes to mimic nature. The waterfalls were stone clad with an imported stone which blended in well with the natural colours and tones of the

sandstone. The planting theme reinforces the natural feel, although it is not all indigenous.

"Natural grasses, such as Swamp Foxtails, provide movement as they sway in the breeze," says Herald. "In other areas we have combined grass like plants such as Dianella's and NZ Flax with the softer foliage of Gardenias. It is important to maintain a balance in any planting theme."

The planting as it grows closer to the house becomes more formal to compliment the architecture and includes shaped Buxus and Lillypilly hedges and ornamental trees such as Cherries and Maples.

Incorporated into the landscape is a perfectly manicured four-hole golf course. The location of each hole carefully designed to include sand bunkers and over water with each flag pole containing the property label.

1 当地桉树
2 瀑布源头
3 傍晚高尔夫球场景观

巴卡花园

Bakka Garden

撰文：Aaron Weingrod & Michael Abrahamson　　图片提供：Aaron Weingrod　　翻译：谷晓瑞

1　入口
2　凉亭
3　西奈休息坐台
4　下沉花园
5　烧烤区
6　水景
7　就餐露台

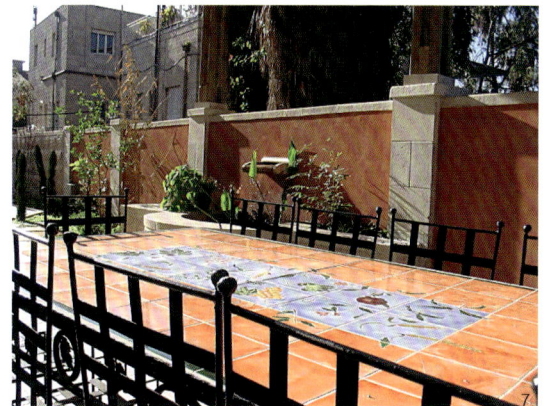

WAA 受邀为巴卡的一个家庭设计花园，该项目所在的地段是耶路撒冷最有发展前景的街区。

客户家中有 5 个孩子，一家人性格开朗，充满艺术气质，他们从一开始就参与到花园的设计之中。花园的设计构思是将其与新家打造为一个整体，一条蜿蜒的小河流过规划后的户外区域。无论是在房屋狭窄的侧面，亦或是房屋的前后，其每个区域都具有独特的用途和设计理念。

设计师根据客户的要求，将该项目规划成 13 个彼此独立而又相连的户外区域。沿着房屋前的车库墙或是在屋后葡萄藤下的角落里设计一排坐位，在铺路时

使用垫石和熔岩穴，这些都是根据每个区域各异的环境和作用而设计的。

走进正门，一条小径向内延伸，拱形通道的支架上爬满了藤蔓，前门的平台一直延伸到圆形地下洞穴区域并与庭院相连（打开前门，一条通向房屋的走廊一直延伸到花园，旁边点缀着一座东方式喷泉），庭院设有一个独立的客人公寓。沿着天然的石阶走上去，就来到了后花园，这里设有木质眺望台、挂在两棵橄榄树间的吊床、一排位于角落的西奈休息坐台、一片可供孩子们嬉戏和掷飞盘的草地、带有凉亭和烧烤区的就餐露台以及一座长满繁花、香气四溢的花园。

1　前园
2　草坪与吊床
3　下沉花园
4　台阶
5　后园
6　喷泉

This garden was designed around a new one family home in Bakka, one of Jerusalem's up and coming neighborhoods.

The clients, a very active and artistic family with 5 children were involved with the garden design from the start. The garden was conceived with the new home as a whole entity, allowing a continuous flow around the house through the creation of many "outdoor rooms". Each of these rooms, whether on the narrow sides of the house or in the front and back, has its own unique use and concept. Working with a rich "wish list" Weingrod-Abrahamson Architects created 13 separate, yet connected outdoor rooms. The use of built- in seating, in the front along the garage wall, or in the back corner under a grapevine; using cushions and a fire pit in the paving—all these created spaces for different atmospheres and functions.

The rooms around this home read as follows—from the front gate: Entrance path and covered vine gazebo, front door platform (opning the front door one sees hallway flooring continuing through the home and into the garden, offset by an Oriental type fountain) continuing round one descends to a below-ground Grotto area with patio, for a separate guest apartment; climbing up natural rock stairs, we reach the back yard with a hardwood meditation platform, a hammock between 2 olive trees, a "Sinai sitting corner" with fire pit, grassy area for running and Frisbee throwing, up to the dining patio with a sukkah / pergola and barbeque area, and a spice garden with a twist!

沙漠一隅 —— 天堂谷

A Piece of the Desert — Paradise Valley Residence

撰文 / 图片提供：Floor 公司　编辑：Pedro F Marcelino　翻译：李沐菲

　　该项目的现代式设计由著名设计师 William Cody 提出，美国棕榈泉市也是由他设计的，但在后来的几十年中，不同的所有者又对其进行了一些与原设计理念不同的改动。该项目位于亚利桑那州最具开发价值的地段，一直面临着拆迁和兴建新的大型住宅的风险。然而幸运的是，目前的所有者意识到了它的价值，并且认为可以通过适当的改造使其能够更好地展现出原设计的精髓以及沙漠景观的风貌。

　　在景观设计中，设计师对该项目的大部分区域进行了修建，包括所有的外部修缮，如新建的游泳池及其相邻的平台、新的车道以及停车场、新的植被以及低电压照明设施等。设计中流露着设计师对中世纪建筑风格的崇尚，保持着一种严谨而又精细的设计理念，同时充分利用了所在区域的自然地理条件来捕捉那些壮观的景色，始终坚持可持续性的理念，保留所在区域的景观特征。

新设施的安装十分精细，以确保其坚固耐用，同时这些设施也增强了住宅空间的立体感。该项目的主体建筑规划中有一个边长为 1.5m 的正方形下层预制住房单元，所有新的硬质景观均根据这一结构的大小而建，并与原有的设计相匹配，以免对原有设计的线条和结构造成破坏。另外，该区域本身也为设计带来了很多挑战：如需要保留住宅后庭原有开放式的感觉；能够与相邻的高尔夫球场相通；需要在确保泳池安全性及私密性的同时，又能欣赏到驼峰山的壮丽景色等。

新建的泳池与主建筑一样采用南北走向，并最大程度地降低了对后庭景观的影响。为了保持视觉的连续性，泳池露台与原有的悬臂式露台位于同一平面，整个泳池都位于地面之上，这种设计对原有斜面以及排水系统的改动程度是最小的。同时，这种将泳池高度提升的设计使其呈现出一种奇妙的缺少边界的效果，

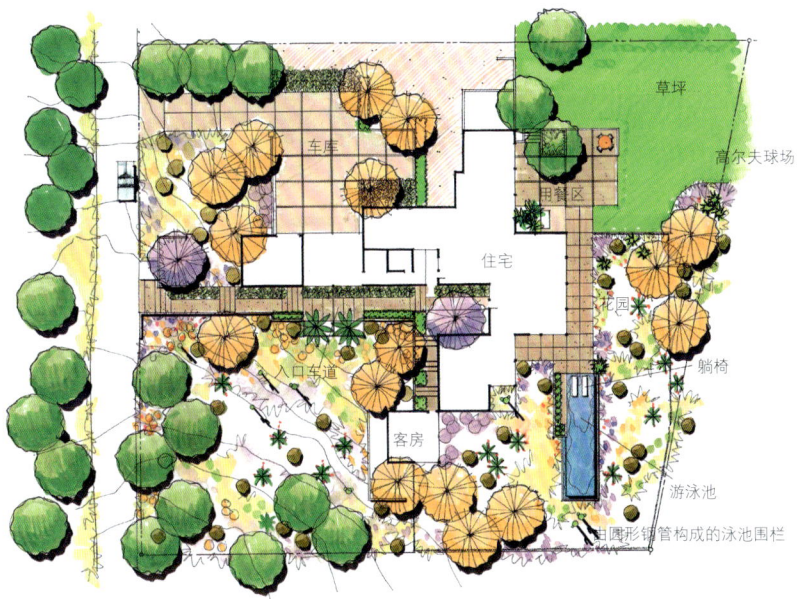

总平面图

227

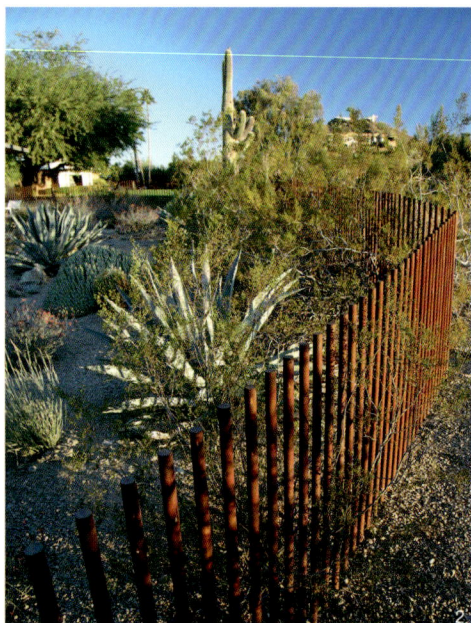

使泳池的三面看起来都具有隐形的效果。泳池宽3m、长12m、高1.5m。在距离泳池3m的地方有一个"玛格丽塔露台"，上面摆放了两个玻璃纤维材质的躺椅，看起来如同漂浮在水面上一样，躺椅在不需要的时候也可以轻易地被移走。

泳池的围栏由圆形的钢管构成，直接固定于地面，而不需要任何水平结构的支撑，这样的设计使其更加通透且不会影响人们欣赏周围的景观。围栏的通透形式以及天然锈蚀的外表面都与当地的沙漠景观相得益彰，丝毫不影响人们欣赏附近的山脉景色以及落日的余晖。为了避免损害植被，每个围栏的位置和安装都十分谨慎，其中古老的墨西哥三齿拉瑞阿的枝条似乎都已经伸出了围栏，这种谨慎的设计也使该项目的景观看起来经久不衰。进入泳池的门也采用了与围栏同样的设计，只是门的高度较低，模仿了本地的建筑风格并与其双开式弹簧门相呼应。这扇门距离泳池露台3m远，既保留了原有的建筑结构，又可以在需要的时

候打开呈90度角，最大限度地扩展了休闲区域的面积。

设计师将泳池所需要的设施系统地安装在一座新墙的后面，这座新墙是为了与这里的一处有着40年历史的残垣相呼应。同时，为了与住宅的建筑线条保持一致，墙体的一端遮蔽泳池设施，另一端向北延伸。许多开小花的树木星星点点地遮挡住了各种设施的连接处，仙人掌果和墨西哥刺木在竖直的墙面上投下了变幻莫测的影子。

独特的车道设计与整个建筑的风格并无任何关联，与整齐的、边长为1.5m的正方形的预制住房单元也无任何相通之处。一系列新建的边长为3m的正方形混凝土板以及混凝土板缝隙之间填充的鹅卵石清晰地界定了行车区域，同时也为客人提供了充足的停车位，而增设的钢板门则将汽车旅馆区与车道分隔开来。重新规划的建筑部分还包括原露天停车区周围的玻璃墙，使之与其他停车场一样用做展示区。

设计师在住宅及泳池露台的周围种植了各种盆栽

植物，通过其抽象的外形、独特的形态以及丰富的纹理来为整个设计增添光彩。汽车旅馆区以及露台区所使用的光滑的鹅卵石也为这一设计增加了现代气息，并将该项目中的不同元素与所在区域紧密地联结在一起。

前庭内较随意地种植了沙漠植被，增加了植被种类，如仙人掌、豆科灌木、墨西哥三齿拉瑞阿以及龙舌兰等。为了丰富其植被种类并保证该项目西、南两个方向的私密性，设计师还种植了一些当地的其他树木和植被。各种植被的组合使这里一年四季都开满各种花朵、具有多样的色彩以及保护性的荫蔽，这里俨然成为了一座植被结构完善的野生植物园。

该项目展示了一个优化的、可持续的设计，同时还保留了原住宅的建筑结构。该项目在尊重原设计理念的前提下，进行了恰当的改进及补充，塑造出了壮观的空间景致，并成功地对居住环境内的外部界限进行了模糊处理。

1　沙漠植物景观
2　漆成红色的栅栏
3　简洁的栅栏
4　住宅与开放的庭院

The modern design of this home was the work of the notable Palm Springs architect William Cody, yet over the past several decades various owners made modifications that ran counter to the original design intent. Located on some of the most valued real estate in Arizona the home was at risk of being demolished to make room for new larger home. Fortunately, the current owners had the vision to see that this home and property could be appropriately renovated and updated to honor the spirit of both the original design and its native desert setting.

The landscape design included all exterior improvements including a new swimming pool and adjacent deck space, new driveway and auto court area, as well as new planting and low voltage lighting renovations over the majority of the property. Complimenting the mid-century architectural style, the landscape design maintains a restrained, yet detailed approach, while taking full advantage of the site's natural topography to capture spectacular views and reintroduces sustainable native landscape palette throughout.

Great care was taken to ensure the new site amenities added to the property did not detract, but instead strengthened the geometric planes that form the spatial volumes of the existing home. The fundamental architectural plan has an underlying design module based on a five foot by five foot square. All new hardscape elements adhere to this module size and align with existing layouts so as not to interfere with the lines and flow of the existing structure. In addition the site itself posed challenges that included the need to preserve the open feeling of the backyard, provide connection to the adjacent golf course and preserve the dramatic views of Camelback Mountain while providing pool security and screening of neighboring residences.

Great care was taken to ensure the new site amenities added to the property did not detract, but instead strengthened the geometric planes that form the spatial volumes of the existing home. The new pool compliments the main north-south orientation of the home, while minimizing the impact to the backyard landscape. For visual continuity, the pool deck elevation was set to match the existing elevation of the cantilevered decking. By keeping the pool entirely above ground, modifications to the existing slopes and drainage flows of the site were minimized. This elevation match allows for a dramatic negative edge pool with a vanishing effect that occurs on three full sides of the pool. The pool itself is ten feet wide by forty feet long with five foot wide planters to either side. The first ten feet of the pool is a shallow "margarita deck" where cast fiberglass chaise lounges seemingly float on the pool's surface by way of an in-ground mounting detail that allows for easy removal of chaises as needed.

1　天堂谷沙漠日落景观
2　从栅栏一侧观赏到的景观

The required pool fence was constructed of round pipe steel, anchored below grade to eliminate the need for horizontal support members and to create a see-thorough effect to adjacent landscape areas. Left in a natural state to rust, the transparent forms of the fence blending into the desert landscape allow for unobstructed views to the adjacent mountains and glowing sunsets. The location and construction of the fencing was carefully orchestrated to maintain all existing specimen plants and cacti, including the ancient creosotes that appear to have grown through the fencing. This careful placement creates an impression of longevity and permanence to the landscape. The pool gates follow the vertical design of the fence utilizing a lower horizontal member to mimic the house's vernacular while supporting its swinging motion. The gate to the north spans the ten foot width of the existing deck without touching it, allowing for preservation of the existing construction, and when needed the gate can be opened 90 degrees to maximize the available exterior entertaining area.

The pool equipment is strategically located behind a new wall that was treated to help match the 40 year old slump-block masonry used in the construction of the home. In homage to the lines of the home, the wall elevation is set at one end to screen the equipment and then transition into the existing grade to the north. Masses of small flowering trees screen unsightly utility connections, and Prickly Pear cacti and Ocotillos cast abstract shadow patterns on the vertical wall surfaces.

The original driveway's curvilinear layout did not match the character of the architecture or bear any relation to the organizing five foot square module. In its place a series of new ten foot square, integrally colored concrete slabs with pebble filled joints were introduced to better define the drive area and provide ample space for guest parking. A new steel plate rolling gate was added to close off the auto court space from the drive. The architectural portion of the remodel

included enclosing of the original open air carport with glass walls to provide a showroom quality to the typical utility-like nature of the common automobile parking space.

Within the planters near the house and around the new pool deck, organized mass plantings of sculptural plants were installed to compliment the architecture through their abstract forms, striking silhouettes, and rich textures. The same smooth pebble-top-dress used in the auto court and rear patio area was used in the planters to compliment the modern look of the home and tie the various elements and areas together throughout the site.

Looser, more informal desert plantings occur in the front and rear yard areas where they augment the existing palette of Saguaro, Mesquite, Creosote and Agaves that were retained in place. Additional native tree and plant materials were added to complement this existing vegetation and help screen views of neighboring residences and roadway to the south and west respectively. The combination of species now found on site offer a seasonal abundance of flowers, color, and protective cover resulting in a textural garden teeming with wildlife.

Project demonstrates a refined and sustainable approach to the preservation of an architecturally significant home. Appropriate upgrades and amenities were implemented while honoring the original design intent. The landscape design creates spectacular spaces and views and successfully blurs the lines between the interior and exterior living environments.

1　游泳池
2　游泳池周围的植物景观

位于门罗的住宅

Residence in Monroe

撰文 / 图片提供：Sawyer/Berson Architecture & Landscape Architecture, LLP　　翻译：武秀伟

该项目位于一座小型半岛上，两条支流在这里交汇。一条曲折的车道穿越一丛橡树林，通向院子入口。其中一条支流的对面，有一座矩形游泳池，蓝沙岩台阶一直延伸到一大片空旷的草地上，直到草地边缘的一条沿水岸铺设的蜿蜒的木板路。具有民俗风情的家具可供人们用餐、休息，种植着各种植物的植栽容器遍布整个场地。茂盛的橡树丛、木兰花、柏树、杜鹃花、山茶花和许多本土植物，为花园增添了生动的色彩。

经过对场地特性的仔细研究并了解了本土的自然环境后，设计师决定在这处独特的滨水区旁建造一座豪宅，因为这里的环境与其设计理念十分吻合。新建的景观可供人们进行户外活动，为业主带来很多新的期盼。同时这里还有很多相互连接的空间，包括通向海湾的小路。

景观设计师和建筑师在最初规划该场地时，景观概念设计和建筑设计同步进行。最后，成功地将建筑与景观融合在一起。

当地的自然环境为项目提供了良好的环境，路易斯安那州耸起的植物园激发了设计师的设计灵感。从房屋到户外空间，再到滨水区的自然过渡，证明了该项设计是成功的。

蜿蜒的河流为场地内的流线设计提供了概念性的指引，包括穿过前方草坪的入口车道，连接后院的人行道和木板路，整体布局使场地显得更开阔。

设计师计划对泳池周围的平台进行多重分割，在这里设置休息区、就餐区和日光浴区。当然，各个区域都要配备相应的家具设施。菜园位于泳池平台的东北部，被上面的柏树丛遮挡住，成为一处私密空间。通往码头的专用通道与平台和菜园在同一条直线上。

木板路横穿后面庭院的环形路，一端与水池平台的休息区相连，另一端与停车场相连。沿着木板路也可以到达楼梯平台、跌水台阶、码头，并可以欣赏到海湾景观。木板路周边的景观与附近的柏树相衬在一起，在水中映出优美的倒影。

木板路与河流间的楼梯平台可供人们欣赏滨水景色。人们可以暂时把船停靠在这里，到岸上游览；也可以乘船穿越河流，到乡村俱乐部去打高尔夫球。

1　黄昏中的居所
2　入口车道
3　入口台阶

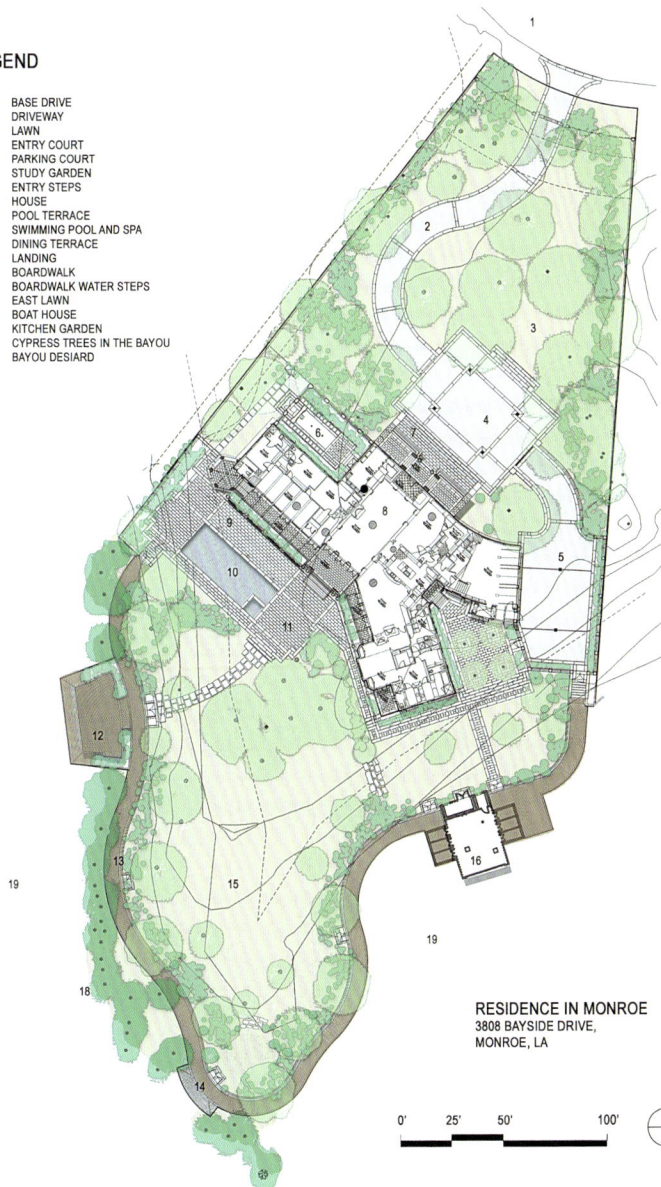

LEGEND

1　BASE DRIVE
2　DRIVEWAY
3　LAWN
4　ENTRY COURT
5　PARKING COURT
6　STUDY GARDEN
7　ENTRY STEPS
8　HOUSE
9　POOL TERRACE
10　SWIMMING POOL AND SPA
11　DINING TERRACE
12　LANDING
13　BOARDWALK
14　BOARDWALK WATER STEPS
15　EAST LAWN
16　BOAT HOUSE
17　KITCHEN GARDEN
18　CYPRESS TREES IN THE BAYOU
19　BAYOU DESIARD

RESIDENCE IN MONROE
3808 BAYSIDE DRIVE,
MONROE, LA

0'　25'　50'　100'

1　影影绰绰的树木临水而立

2　从就餐区可以看到草坪、木板桥、
　　过渡平台和远处的河流

This landscape is on a small peninsula at the convergence of two bayous. A winding driveway leads through an oak grove to an entry court and new house. Facing the bayou is a rectangular pool and bluestone terraces step down to a sweeping lawn that ends at a serpentine boardwalk along the water's edge. Custom furniture, for dining and relaxation, and planters and plantings are placed throughout the site. Lush planting of oak, magnolia and cypress together with azaleas, camellias and a collection of native plants provide the living palette for the garden.

The successful juxtaposition of the extensive site program onto this uniquely configured two acre waterfront setting was the result of careful attention to the existing site attributes; an understanding of the vernacular cultural and natural environment; and conformance to a clear design concept. The built landscape accommodates the outdoor lifestyle and expectations of the owner's family providing a broad range of linked spaces of varying scale, including full access to the bayou.

Initial planning to site the house was a close collaboration between the landscape architect and the architect where conceptual design work for the landscape, and the house, was a simultaneous progression. The result was a successful blurring of the line where architecture meets landscape without precluding opportunities for either.

The landscape provides the base for the house that was inspired by the raised plantation cottages of Louisiana. The unencumbered transitional sequences from the house to the outdoor spaces to the waterfront mark the primary success of this project design.

The meandering bayou provided the conceptual direction for the circulation design within the site; including the entry drive through the front lawn, connecting walkways at the rear yard, and the boardwalk. This layout device resulted in the illusion of a larger site.

Programmed spaces within the pool terrace accommodate mixed uses including the seating area, dining area, and the sun lounge area. The furniture was also designed to accommodate these uses. Conversely, the kitchen garden is located to the northeast of the pool terrace, and functions as an intimate space that is shaded by the overhead bosque of cypress trees. This terrace is aligned between the kitchen and the dedicated path to the boat house pier.

The boardwalk was designed to traverse the perimeter loop of the rear yard bordering the bayou, with linkage to the pool terrace seating area at one end, and the parking court at the other end. It also provides access to events along the way including the landing, water steps, the boathouse, and the spectacular views of the bayou. Views along a portion of the boardwalk are framed by the adjacent foreground of cypress trees growing in the water.

The furnished landing between the boardwalk and the bayou is the primary location for the family to enjoy passive pleasures along the waterfront. It is also the place to temporarily dock the family boat for day excursions on the bayou, or to just cross it to play golf at the country club.

1　蜿蜒的木板路
2　泳池台阶休息处——
　　从远处可以看到东边的草坪和海湾
3　泳池边的坐椅
4　休憩门廊
5　泳池边的就餐露台